SPACE, TIME, AND MOTION
A Philosophical Introduction

SPACE, TIME, AND MOTION
A Philosophical Introduction

WESLEY C. SALMON
University of Arizona

UNIVERSITY OF MINNESOTA PRESS
Minneapolis

Copyright © 1980 by the University of Minnesota.
All rights reserved.
Published by the University of Minnesota Press,
2037 University Avenue Southeast,
Minneapolis, Minnesota 55414

2nd edition, revised, second printing, 1982

Library of Congress Cataloging in Publication Data

Salmon, Wesley C

  Space, time, and motion.

  Bibliography: p.
  Includes index.
  1. Space and time. 2. Motion. 3. Physics—
Philosophy. I. Title.
QC173.59.S65S24   1980   530.1'1   80-18423
ISBN 0-8166-1044-5 (pbk.)

The University of Minnesota
is an equal-opportunity
educator and employer.

TO ADOLF GRÜNBAUM
who has helped me so much
in so many ways

CONTENTS

PREFACE TO THE FIRST EDITION ... ix

PREFACE TO THE SECOND EDITION ... xi

Chapter One
THE TWIN SISTERS: PHILOSOPHY AND GEOMETRY ... 1

Chapter Two
A CONTEMPORARY LOOK AT ZENO'S PARADOXES ... 31

Chapter Three
A TRIP ON EINSTEIN'S TRAIN ... 69

Chapter Four
CLOCKS AND SIMULTANEITY IN SPECIAL RELATIVITY
or WHICH TWIN HAS THE TIMEX? ... 93

EPILOGUE ... 128

NOTES ... 131

BIBLIOGRAPHY ... 141

INDEX ... 153

PREFACE TO THE FIRST EDITION

This book is intended as an invitation to the philosophy of space, time, and motion—a field that has formed a central core of natural philosophy for some twenty-five centuries. It is designed as a sort of sampler, being neither a systematic survey of problems and points of view, nor (although the topics come from many different periods) a comprehensive historical account. It treats, instead, an assortment of issues, at an extremely elementary level, that concern people working in this area. I hope that discussions of such topics as the nature and implications of non-Euclidean geometry, ancient and modern versions of Zeno's paradoxes, and the features of time as revealed by special relativity (including the famous "twin paradox") will prove fascinating enough to motivate the reader to pursue these ideas in more systematic and advanced texts.

The prerequisites for this book are very modest. Chapter 1, "The Twin Sisters: Philosophy and Geometry," should be intelligible to anyone who has studied geometry in high school. Chapter 2, "A Contemporary Look at Zeno's Paradoxes," presupposes no specific background knowledge, although a small acquaintance with differential calculus, such as is offered in many high schools nowadays, might be helpful. Anyone feeling the need for such background can find it easily in Chapters 6-8 of Courant and Robbins, *What is Mathematics?* (Oxford University Press, 1941). Some familiarity with set theory might also be helpful; those portions that bear especially upon the topics of Chapter 2 are presented quite concisely in the Appendix of my anthology, *Zeno's Paradoxes* (Bobbs-Merrill, 1970). Chapter 4, "Clocks and Simultaneity in Special Relativity," presupposes a slight knowledge of the special theory of relativity; the basic prerequisites are contained in Chapter 3, "A Trip on Einstein's Train," which uses no mathematics beyond the high school level.

For a deeper and more systematic introduction (as opposed to an "invitation") one could hardly improve upon the works of Hans Reichenbach, especially his superb classic, *The Philosophy of Space and Time* (Dover Publications, 1958). Attention should also be called to his posthumous book, *The Direction of Time* (University of California Press, 1956, hardcover; 1971 paperback) which raises many issues not even mentioned in this book. For more recent discussion, elaborations, and extensions, often at a more technical level, the works of Adolf Grünbaum deserve special mention—especially the new and greatly expanded second edition of his *Philosophic Problems of Space and Time* (Reidel, 1973). Additional readings are listed at the end of each chapter of the present volume.

I should like to take this occasion to thank the Bobbs-Merrill Company, *Scientific American,* and the Ohio State University Press for permission to quote or paraphrase some of my previous publications. Specific acknowledgements are given at the appropriate places in the text. I should also like to express thanks to my wife, Dr. Merrilee Salmon, and to three anonymous referees engaged by the Dickenson Publishing Company, for reading the manuscript and making extremely helpful comments, suggestions, and corrections.

I should also like to thank Dr. Gerald Moore, Physics and Optical Sciences, University of Arizona, for help in checking my computations in Chapters 3 and 4, and Mrs. Alexis Ahmad for her artistic execution of the figures.

In presenting the discussions that follow, I am deeply conscious of my enormous intellectual debts to Hans Reichenbach and Adolf Grünbaum. Their contributions to this area have been largely instrumental in making twentieth-century philosophy of space and time the fascinating and challenging field that it is.

W. C. S.

PREFACE TO THE SECOND EDITION

I am grateful to the University of Minnesota Press for making it possible to bring out a second edition of this book. The only major alteration is the addition of the annotated bibliography which begins on page 141. I should like to thank Lindsay Waters of the University of Minnesota Press for his kind help and cooperation; Russ Vogt of Media and Instructional Services, University of Arizona, for a new rendition of figure 1, chapter 3; and Alan Wachtel of Stanford University for pointing out a number of typographical errors in the first edition.

W.C.S.

Chapter One

THE TWIN SISTERS: PHILOSOPHY AND GEOMETRY

Almost everyone, nowadays, is aware of the intimate relations between the sciences and the various branches of mathematics. The mathematical character of modern physics is a commonplace, and the uses of mathematics in the biological and social sciences are widely publicized. You might think philosophy, in contrast to the sciences, would bear no especially significant relationship to mathematics—geometry in particular—but such a view would be quite mistaken. To be sure, the relationship is not of the same type as you find between mathematics and the empirical sciences, where some branch of mathematics is a useful, powerful, perhaps even indispensable, tool for the science in question. In the case of philosophy, the relationship goes far back, to a time long before there were any other sciences besides mathematics worthy of the name. It is a very deep relationship, representing one of the most profound and fruitful interactions between any two disciplines in the entire history of human thought. This is a rather strong statement, but I hope to justify it by sketching some of the high points of the story of this relationship.

Geometry and philosophy were born together at the same time, in the same place, and indeed, they had the same father. They are more like twin sisters than servant and master. The Greek philosopher, Thales of Miletus, who flourished around 600 B.C., made a trip to Egypt, where he learned something of the art of the surveyors, and of the geometrical knowledge they had accumulated. He brought this knowledge back to Greece, but he also introduced a significant change of viewpoint. Instead of regarding geometrical truths simply as rules of thumb, or practical guides furnished by experience, he regarded them as propositions which could actually be proved. This transformed geometry into a genuinely mathematical discipline. Thales is credited, for instance, with proving the theorem (among others) that the base angles of an isosceles triangle are equal to one another. At the same time, he is usually cited in the history books as the first important philosopher. We do not know too much about his philosophical work, for little of his writing has survived, but it appears that he believed that all things are composed of water and all things are full of gods. Mathematics seems to have had a more auspicious beginning than philosophy, but however that may be, they did come into being together with the work of Thales.

During the next three centuries both geometry and philosophy flourished; for example, the mathematician Pythagoras and the philosophers Plato and Aristotle made their monumental contributions in this period. Pythagoras, incidentally, is famed as a philosopher as well as a mathematician; he stands with Descartes, Leibniz, and Russell as a philosopher whose contributions to mathematics would assure him immortality on that basis alone. By around 300 B.C. geometry was so well developed that Euclid was able to write his epoch-making work, *The Elements,* in which he reduced the whole of geometrical science to an axiomatic form in which all of the propositions (theorems) are deduced from a very small number of starting assumptions (axioms and postulates). This placed geometry in a unique position. It was by far the most thoroughly developed and most highly perfected science that existed in antiquity; indeed, I think it is fair to say, until the publication of Isaac Newton's *Principia* in 1686—just about two thousand years later—no other science existed which had an equal degree of development and perfection.[1]* The nearest rival is perhaps astronomy, but it was little more than a branch of applied geometry before Newton supplied physical explanations for the motions of the heavenly bodies.

Shortly before the time of Euclid, Plato established his Academy, a famous ancient school of philosophy. It is said that there was a sign over the door which read, "Let no one enter here unless he knows geometry." After Euclid had written *The Elements*, Plato's successors at the Academy are said to have changed the sign to read, "Let no one enter here unless he knows Euclid." Now, this might strike you as the kind of dark saying that philosophers are supposed to be famous for, but actually Plato had a very good reason for his attitude toward geometry. For him, geometry held the key to philosophical knowledge and truth—it held the key to the understanding of

---

*Notes to each chapter appear at the end of the book.

reality. Philosophers generally are interested in questions concerning the nature and the foundations of human knowledge. They want to know what knowledge is, how it is acquired, and on what basis it rests. Plato certainly was deeply interested in questions of this sort.

When Plato looked around for an example of human knowledge, geometry must have struck him as the outstanding candidate, for it was by far the most perfect instance of knowledge available at the time. It was evident to him that geometry was a science of pure reason; it was not the empirical science of the Egyptians who had learned by experience, for instance, that a triangle with sides of three, four, and five units, respectively, must contain a right angle. By taking a rope twelve units long, and marking off the units, they could use this knowledge to provide a practical method for constructing right angles—a task of considerable importance in surveying and building. The geometry Plato was thinking about was much like that of Euclid, where abstract propositions about abstract entities—perfect straight lines, perfect circles, perfect triangles—were proved by pure reason. To Plato this indicated that man's *reason* is capable of achieving knowledge of the most important sort; indeed, he believed the abstract geometrical entities to be more real than ordinary physical objects which exemplify these figures only imperfectly. Geometry thus constituted the best example of scientific knowledge, and in Plato's view it provided the keys both to the nature of human knowledge and to the nature of ultimate reality.

This attitude was by no means confined to Plato. Throughout hundreds of years, philosophers have held that pure reason does provide us with knowledge of the world, and that it is the best source of knowledge we have. It is far superior to our senses, which are quite capable of leading us astray. Thus, when René Descartes (1596-1650) ushered in the modern period of philosophy by wholesale doubting of the reliability of sense experience, he was not doing anything new. When, in his *Meditations,* he argued that we could never be absolutely sure of the pronouncement of the senses because we could never know for certain that we are not dreaming, Descartes was simply giving forceful expression to an idea that was familiar from before the time of Plato; namely, that our senses are subject to illusion, and they can therefore deceive us. And when he looked to geometry for a way out of the difficulty, arguing that geometrical properties (extension) represent the essence of matter, he was undertaking a program he was neither the first nor the last to attempt. He was also, incidentally, squarely in the tradition of Thales, for Descartes's invention of analytic geometry was an extraordinary mathematical contribution.

The doctrine that geometry provides useful knowledge of the physical world via pure reason was given its clearest formulation by the eighteenth-century philosopher, Immanuel Kant, who said that the propositions of geometry are *synthetic a priori* truths. By *a priori* he meant that they can be deduced by pure reason from postulates that are apparently self-evident; we do not have to perform observations and experiments in order to prove geometrical theorems. In addition, geometrical propositions are *synthetic*, which means that they provide information about the physical world in which we live—information which is clearly useful in such enterprises as surveying, navigation, architecture, engineering, and the natural sciences. In other words, geometry provides us knowledge of the actual structure of the space in which we live and move, and of the spatial relations among the objects that we meet with in everyday life.

This, then, is the picture of scientific knowledge that emerged from more than two millenia of contemplating the example provided by geometry. Knowledge that can be used to understand, predict, and control the happenings in our world can be established by purely logical demonstration. To be sure, the kinds of observations made by the Egyptian surveyors may *suggest* geometrical theorems to us, but they do not enter into the geometrical *proofs* in any way. Such observations are as irrelevant as the figures we draw on a chalkboard to the demonstrations of geometrical theorems. Empirical observations may serve a heuristic function, but they have no bearing upon the proofs.

Demonstrations must, of course, start somewhere, and this is where the *postulates* come in. The postulates are, so to speak, the basic premises for all of the geometrical deductions.[2] This naturally leads us to question their status. These basic propositions, from which all of the theorems of geometry are supposed to follow, were long regarded as self-evident propositions. You can see that they must be true by just contemplating them—no one in his right mind could doubt them. Even John Stuart Mill, in his classic essay "On Liberty" (written several decades after the discovery of non-Euclidean geometry)—an essay devoted to defense of free discussion of all sorts of issues—can find no value in disputing the postulates of geometry! Thus, scientific knowledge is seen as a series of deductions from self-evident premises.

The view that the postulates of geometry are self-evident truths, though widely held, was not universally shared. As a matter of fact, we are not quite sure how Euclid stood on this point. He seems to have had no doubts about the first four of his postulates:

P-1. A straight line can be drawn between any two points.
P-2. A finite straight line can be extended continuously in a straight line.

> P-3. A circle can be drawn with any center and any radius.
>
> P-4. All right angles are equal to one another.

But there was, in addition, the infamous fifth postulate—the parallel postulate:

> P-5. Given a straight line and a point not on that line, there is one and only one line through that point parallel to the given line.[3]

Euclid proved the first twenty-six propositions at the beginning of his *Elements* before he made any use at all of the fifth postulate. It looks a bit as if he wanted to prove all that he could without having to employ the parallel postulate; perhaps he regarded it as a little dubious, unlike the other four.

Whatever Euclid may have thought, subsequent geometers certainly regarded the fifth postulate as less self-evident than the other four. It was not that they doubted its truth; it was rather that the fifth postulate was more complicated than the others, so that its truth may not have been equally obvious. For more than two millenia mathematicians engaged in futile efforts to prove the parallel postulate. All were unsuccessful. Either they committed some logical fallacy, or they substituted an assumption which is just as difficult to justify as Euclid's fifth postulate. Some of the alternative assumptions which enable one to derive the parallel postulate are quite interesting. For example, it is sufficient to assume that the sum of the angles of the triangle is equal to two right angles, or that there are triangles of the same shape (similar triangles) but not of the same size, or that a line which is everywhere equidistant from a straight line is itself a straight line. Attempts to prove the parallel postulate on the basis of such assumptions are, of course, question begging, for these assumptions are equivalent to the parallel postulate itself.

One particular attempt to prove the fifth postulate deserves special mention. In 1733, Girolamo Saccheri, an Italian Jesuit, attempted to prove the parallel postulate by assuming it to be false, and then deducing an absurdity. This form of argument is known as *reductio ad absurdum,* and it is perfectly valid; in mathematics it is sometimes called "indirect proof." Saccheri's task split naturally into two parts. In the first place, one can deny the parallel postulate by maintaining that parallel lines do not exist at all. On the basis of this assumption Saccheri did succeed in deriving a contradiction, for the first four postulates *do* imply that there is at least one line through the given point parallel to the given line. In the second place, one can deny the parallel postulate by asserting that there is more than one line through the given point parallel to the given line. Saccheri satisfied himself that he had deduced a contradiction from this assumption as well, but in fact he did no such thing. What he actually succeeded in doing was deducing some interesting theorems of what later came to be known as non-Euclidean geometry. Unfortunately, he mistook one of the stranger ones for an absurdity. He was, therefore, the unwitting discoverer of non-Euclidean geometry, even though he died believing he had "cleansed Euclid of every blemish."[4]

Saccheri did not know what he had done, but near the beginning of the nineteenth century three famous mathematicians, Carl Friedrich Gauss, Johann Bolyai, and Nikolai Ivanovich Lobachevski, working on the problem of the parallels, came to the conclusion that it is possible to assume that the parallel postulate of Euclid is false without getting into any absurdity or contradiction. In fact, they realized, it is possible to adopt Euclid's first four postulates while denying the fifth one (by asserting the existence of more than one parallel), and to develop a perfectly consistent non-Euclidean geometry on that basis. Gauss, who was quite possibly the greatest mathematician of all time, was the first to make the discovery, but he was so reluctant to engage in the kinds of senseless dispute he felt would result from publication that he did not publish his discovery until much later. In the meantime, about 1820, the same results were established more or less simultaneously, and certainly independently of one another, by Bolyai and Lobachevski. They denied Euclid's fifth postulate by saying that, instead of one parallel, there are many parallel lines, and on this basis they developed a new geometry.

A number of years later, around the middle of the nineteenth century, the mathematician Georg Friedrich Bernhard Riemann discovered that it is possible, if one tinkers a bit with the first four postulates, to develop another type of non-Euclidean geometry on the basis of a postulate that denies the existence of parallels altogether. He carried out this program successfully. Thus, at about the time Kant died, Gauss was secretly working out the details of a non-Euclidean geometry, and about twenty years later Bolyai and Lobachevski published their versions of such a geometry. Fifty years after Kant's death, Riemann elaborated a second form of non-Euclidean geometry. At this point, three distinct types of geometry are available: the geometry of Euclid (one parallel), the geometry of Bolyai and Lobachevski (many parallels), and the geometry of Riemann (no parallels). In addition, Riemann had worked out a generalized system of geometry in which the foregoing three fit as special cases.

In order to get an intuitive feel for these geometries, it will be a good idea to see how they can be realized. If we confine attention to the two-dimensional case, it is easy to see some of the important distinguishing characteristics of these types of geometry, for each is exemplified by a two-dimensional surface of a particular sort. As we all know, a two-dimensional surface like the surface of a chalkboard represents part of the two-dimensional Euclidean plane. Lines, circles, triangles, and so on, in the ideal flat surface satisfy the relations laid down for two-dimensional Euclidean geometry.

A different sort of geometry is illustrated by the surface of a sphere. The sphere itself is a three-dimensional solid object in three-dimensional Euclidean space, but its surface is two-dimensional, as is indicated by the fact that points on the surface of the earth (which is approximately spherical) can be unambiguously located by two coordinates (longitude and latitude). If we look at this surface in the same way as we regard the surface of the chalkboard when we do ordinary plane Euclidean geometry, we can use it to illustrate one form of non-Euclidean geometry—namely, the Riemannian geometry of no parallels. In order to see this we must, first of all, be quite clear as to what we mean by a "straight line" on the surface of a sphere.

At first blush it looks as if there are no straight lines on the surface of the sphere; if we try to join two distinct points on the surface by a straight line, that line will not lie on the surface but will depart from it. For instance, to join the north and south poles of the earth (thinking of it as if it were a perfect sphere—which, of course, it is not) we get the axis of rotation which goes through the center of the earth and does not remain on its surface. But, by a suitable definition, we can introduce the concept of straight lines which do lie on the surface. The definition we need is simply the customary one according to which a straight line is the shortest distance between two points—but the line must remain on the surface. It cannot bore through the sphere, nor can it leave the sphere and go out into space—it must stay right on the surface. Defined in this way, the straight lines (often called "geodesics") on the sphere turn out to be the so-called "great circles."

Thinking again of the earth as a sphere, we see that one of these great circles is the equator, and all the lines of longitude are also great circles. Indeed, any plane that passes through the center of the sphere intersects the surface of the sphere in a great circle. You can test these claims by stretching a string between points on the surface of your globe. The lines of latitude, as you will see, are definitely not straight lines; they are circles which are *not* great circles. They are not shortest paths between the points that lie in them. Consequently, if you want to travel on the surface of the earth between two distant points on the same latitude, you take what is called a "great circle route," which is familiar in connection with intercontinental air travel. This route does not follow the same latitude, but rather a great circle. If, for instance, you want to fly from Los Angeles to Jerusalem, both cities being at roughly 30° north latitude, your great circle route would take you over Greenland at about 60° north latitude (halfway to the North Pole from 30°), but that is the shortest route (see Figure 1). If you have any doubt, test it with your stretched string.

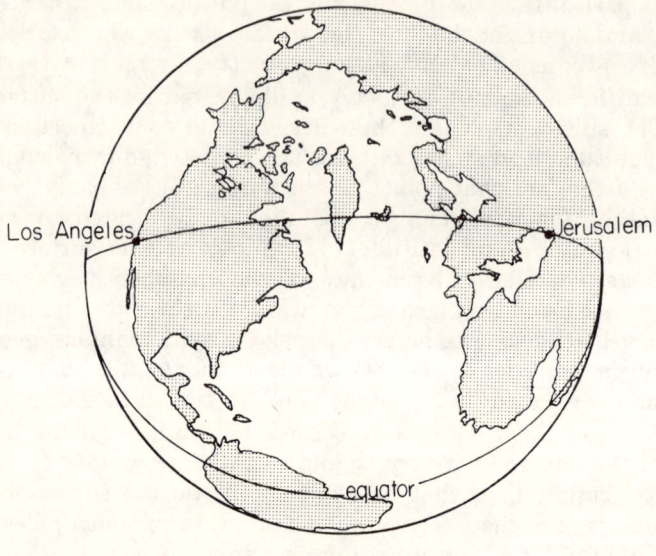

FIGURE 1.

Another way to see that lines of latitude are not shortest paths is to consider two opposite points on a line of latitude drawn close to the North Pole. It is obvious that to get from one point to the other you would not follow the circle around the pole; you would go straight across the pole instead.

With this understanding of what qualifies as a straight line, we can ask questions about parallel lines, triangles, and circles. From here on, we shall use the unqualified word "line" to refer to straight lines; for example, when we speak of parallel lines, these are to be understood to be *straight* lines that do not intersect one another. First of all, it should be obvious that there are no parallel lines on the surface of the sphere because all great circles intersect one another. The lines of longitude intersect at the two poles, the equator intersects the lines of longitude, and any other great circle that you draw is going to intersect the lines of longitude, the equator, and any other great circles that there might be. This is, indeed, a geometry of no parallels: given a straight line and a point not on that line there is no straight line through that point which does not intersect the given line.

Circles and triangles on the surface of the sphere differ remarkably from circles and triangles in the Euclidean plane. Consider a large triangle ABC (see Figure 2) whose base line is along the equator and whose sides run from the equator to the north pole along lines of longitude. The lines of longitude intersect the equator in right angles, so that the two base angles at B and C are both right angles. The sum of the angles of triangle ABC is, consequently, greater than two right angles by the amount of the angle at A. We have all learned in Euclidean geometry that the sum of the angles of a triangle is two right angles, so here we have one of the basic distinctions between this particular type of geometry and the familiar Euclidean geometry we studied in high school. The sum of the angles of the triangle in Euclidean geometry is always 180°, whereas, in the non-Euclidean geometry of the sphere the angular sum is greater than 180°. Moreover, on the sphere the sum of the angles is not constant but depends rather upon the size of the triangle. If you draw a very tiny triangle (DEF, Figure 2), the sum of the angles will be very nearly 180°; if it is small enough the angular sum will be indistinguishable for practical purposes from 180°

FIGURE 2.

Consider next a circle that is drawn on the surface of the sphere (Figure 3). The circle is simply a locus of points equidistant from a given center P. We can, of course, draw a straight line QR through the center; QR is a diameter. We can now ask about the relationship between the circumference and the diameter of this circle.

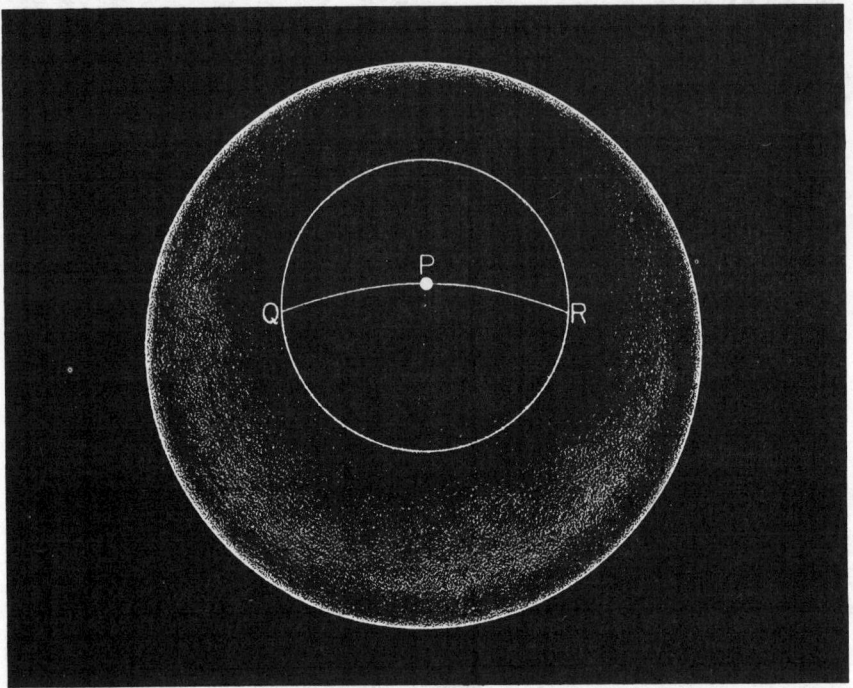

FIGURE 3.

Once again, in Euclidean geometry, we have learned that the ratio of the circumference to the diameter c/d is the number $\pi(= 3.1415 \ldots)$. In the geometry of the sphere, however, this ratio is smaller than π, roughly because the diameter QR has to bend a little bit to get from one side of the circle to the other. You can see this very clearly by considering the equator as a circle. As a matter of fact, it is a great circle, and so it is both a circle and a straight line, paradoxical as that might seem! Consider a diameter of the equator; it is simply a line of longitude. The line of longitude extending from the equator on one side, through the pole, to the equator on the other side, is obviously just half the length of the equator itself; consequently, in that case the ratio $c/d = 2$. Again, with the ratio c/d, as with the angular sum of the triangle, the value is not constant but depends upon the size of the figure. If we have a very large circle, the discrepancy between c/d and π is rather large; in a small circle the ratio will be very nearly equal to π. As the circle gets very, very tiny, the ratio becomes practically indistinguishable from π.

The foregoing remarks have brought out a couple of the most outstanding features of the non-Euclidean geometry which is represented by the surface of the sphere. The geometry of many parallels, developed by Bolyai and Lobachevski, is exemplified by a surface

THE TWIN SISTERS: PHILOSOPHY AND GEOMETRY

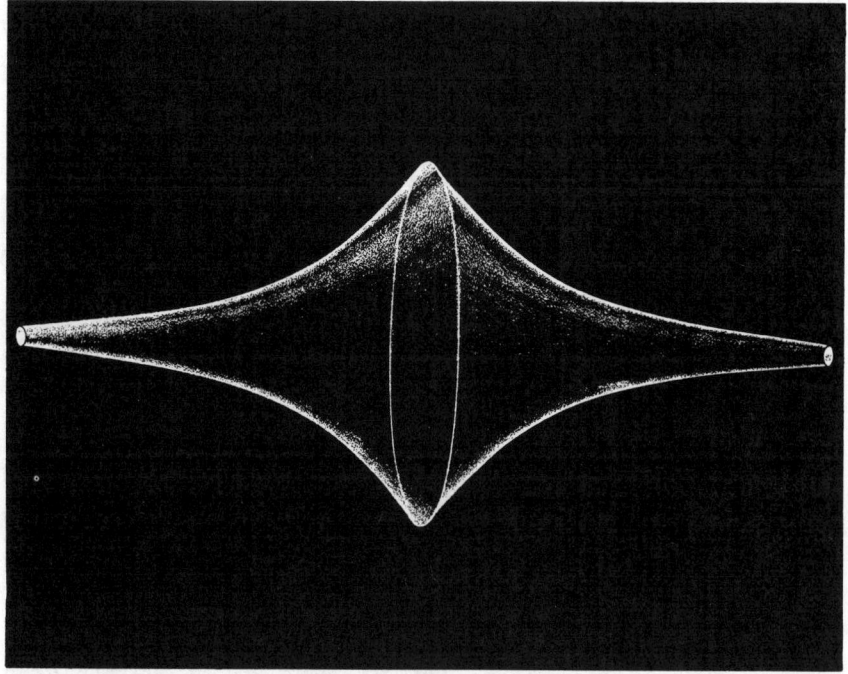

FIGURE 4A.   Pseudo-sphere.

which is known as a pseudo-sphere (Figure 4A). The saddle surface (Figure 4B) provides a useful approximate representation of that type of geometry.[5] The important thing about surfaces like the pseudo-sphere and the saddle is that, unlike the sphere, if you look at one cross section of the surface (front to back) it curves down; if you look at a perpendicular cross section (side to side), it curves up. The sphere, of course, curves in the same direction regardless of which cross section you examine. This is the basis for saying that the surface of the sphere has positive curvature while the surface of the pseudo-sphere has negative curvature.[6]

    In dealing with the geometry of the saddle surface (or pseudo-sphere) we adopt the same meaning as before for "straight line"— namely, the shortest distance between two points. In Figure 5 we have a picture of a saddle surface with a straight line L and a point P above, with two straight lines through the point P. These two solid lines are both parallel to the original given straight line L, for no matter how far they are extended in either direction they will not intersect L. They approach L (in opposite directions), but they never meet L. Moreover, they both go through the point P, so here we see how the saddle surface realizes the condition of this type of non-Euclidean geometry: given a straight line and a point not on that line,

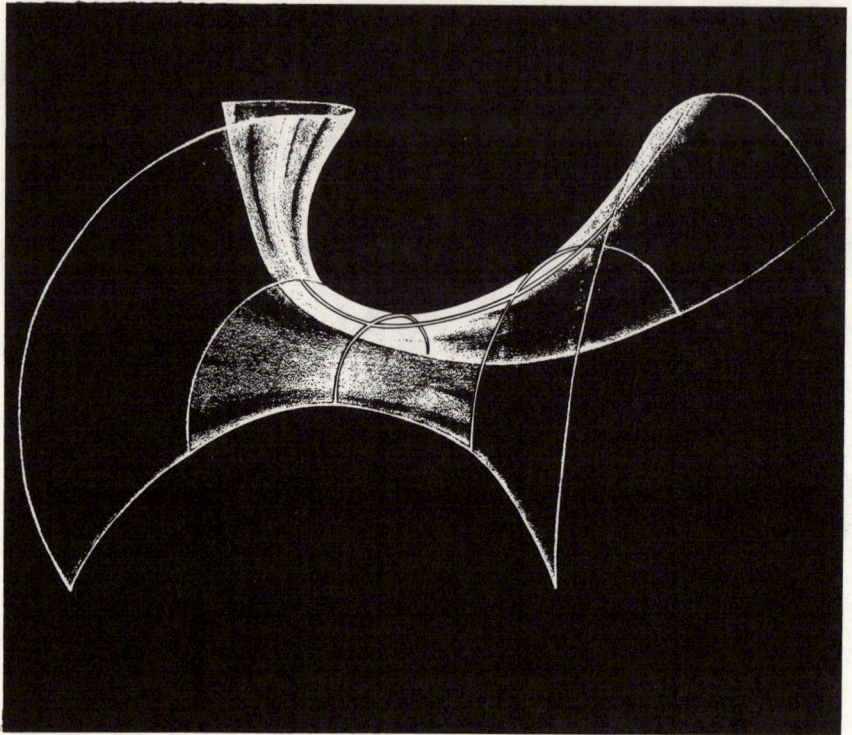

FIGURE 4B. Saddle.

there is more than one straight line through that point parallel to the given line.

In Figure 6 we have a triangle drawn upon the saddle surface. The triangle appears to be concave, but the sides of the triangle are indeed straight lines—that is, shortest paths on this non-Euclidean surface. You can see quite clearly that the sum of the angles of that kind of triangle will be less than the sum of the angles on the flat Euclidean surface. Thus, on the saddle-type surface, the sum of the angles of a triangle is less than 180°. As in the case of the sphere, the sum of the angles is not constant, but depends upon the size of the triangle. The larger the triangle, the larger the discrepancy between the angular sum and 180°; the smaller the triangle, the smaller the discrepancy. In a very tiny triangle, the discrepancy is not noticeable at all.

Now, although it is hard to illustrate this point with a picture, it is nevertheless true that the ratio c/d of the circumference to the diameter of a circle in the saddle surface differs from π, the Euclidean value. In this case, c/d is greater than π, instead of less as in the case of the spherical surface. If you happen to have a saddle (or a Pringle

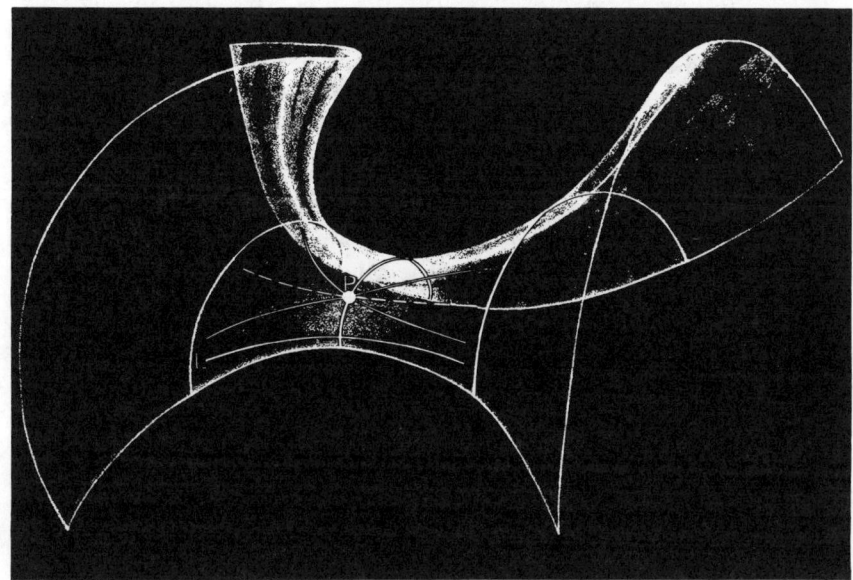

FIGURE 5.

"Newfangled Potato Chip") you might try verifying this fact using a string, a string compass, and a ruler. Again, the discrepancy depends upon the size of the figure.

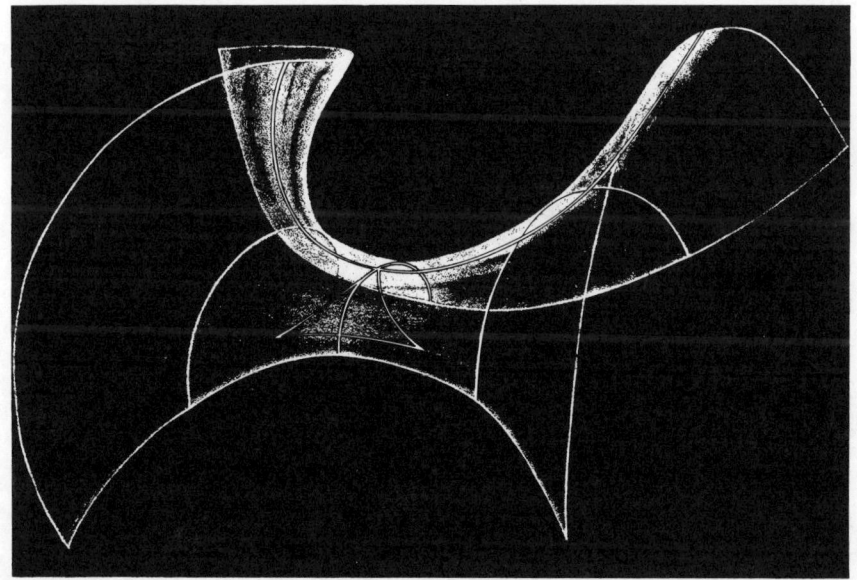

FIGURE 6.

We have now presented several major features of the three types of geometry; they are summarized in Table 1. But what sort of moral are we to draw from all of this? After all, for well over 2000 years there was one and only one geometry, and it was universally accepted as *the* true geometry. It evidently did not occur to anyone to suppose seriously that there were others, different from Euclid's geometry, which might be equally acceptable. Then, a mere century and a half ago, the non-Euclidean geometries were developed. The situation changed markedly. Now, there are several different kinds of geometry to choose among, and the question naturally arises as to which is correct. Each type can be elaborated in a purely formal and axiomatic fashion. Each can be formally generalized to three dimensions, just as Euclidean plane geometry can be generalized to Euclidean solid geometry. Furthermore, from a purely formal standpoint, all of these geometries are on a par. While it is impossible to provide an unconditional proof that the non-Euclidean geometries are free from contradiction, no such proof can be given for Euclidean geometry either. But it is possible to prove relative consistency: if any one of these geometries contains a contradiction, so does every other one. In particular, if non-Euclidean geometry is inconsistent, then so is Euclidean geometry. With this discovery the *logically* privileged status of Euclidean geometry was forever destroyed.

TABLE 1

	Riemann	Euclid	Bolyai-Lobachevski
Parallels	Zero	One	Many
Surface	Sphere	Plane	Pseudosphere (or saddle as a good approximation)
Curvature	Positive	Zero	Negative
Angular sum for triangles	$> 180°$ depends on size of triangle	$= 180°$ independent of size of triangle	$< 180°$ depends on size of triangle
Ratio of circumference to diameter of circle	$< \pi$ depends on size of circle	$= \pi$ independent of size of circle	$> \pi$ depends on size of circle

THE TWIN SISTERS: PHILOSOPHY AND GEOMETRY

If there is no logical basis for selecting a particular geometry, the next question is obviously, "Which of these geometries correctly describes the physical space of our universe?" Gauss, himself, saw this question clearly and performed an experiment in an attempt to answer it. He went to the tops of three mountains in the Alps, and he sighted from one mountaintop to another to lay out a triangle with the mountaintops as the three vertices. The sides of these triangles were defined by the paths of the light rays along which he sighted. This was *not* a triangle on the surface of the earth; the light rays go through space near the earth's surface, but the triangle touches the earth only at its three vertices. He made careful measurements of the angles of this triangle but, within the experimental accuracy available in that case, he could find no discrepancy between his results and 180°. We are not terribly surprised to learn that he was not able to find any disparity between triangles in actual space and Euclidean triangles, because his terrestrial triangle was, in terms of the total size of the universe, an exceedingly small one. We recall the point made above: in each type of non-Euclidean geometry the difference between the angular sum of a triangle and two right angles is a function of the size of the triangle—the smaller the triangle the smaller the difference. Even if physical space were non-Euclidean on a cosmic scale, the angular sum of Gauss's triangle would be very nearly two right angles. But whether the outcome is surprising or not, Gauss's experiment was philosophically important because of the direct way in which it confronts the problem of the geometrical structure of physical space.

Looking at Gauss's experiment, we may be tempted to suppose that the question of which geometry correctly describes physical space admits of a fairly straightforward empirical answer. If we make more and more precise measurements of larger and larger triangles, and if the results were always the same—no detectable difference from 180°—we would seem justified in saying that, within the accuracy of our methods, physical space reveals no departure from Euclidean relations. But we must consider what would happen if the results were different—if measurement of some triangles were to yield a distinct discrepancy from two right angles. Would we not be required to conclude that the geometry of physical space is non-Euclidean? Unfortunately, a simple unqualified affirmative answer is not warranted.

To illustrate this point, imagine what might have happened if Gauss had obtained a different outcome from the measurement of his Alpine triangle. Suppose, for example, that his very careful measurements had produced an angular sum distinctly below two right angles. What then? The normal reaction at this point would probably be to say, "Something has gone wrong; maybe this figure is not a triangle after all—perhaps the light rays along which we sighted did not travel in straight lines." If the figure had curved sides rather than straight ones there would be no conflict with Euclidean geometry in saying that its angular sum is less than two right angles. It would then be necessary to investigate further and find out whether we were indeed measuring the angles of a triangle, or whether we were dealing with some curvilinear figure instead.

Let us see what might happen as we pursue this question. Suppose that somehow or other we take meter sticks and measure the paths of the light rays between the various mountaintops. And suppose we find that the paths of the light rays are in fact shorter, as measured by our meter sticks, than any other paths connecting these peaks. What then? It would still be open to someone to object, "After all, it could be that strange forces are affecting our measuring rods—perturbations which make them change their size or shape as they are moved from place to place." Again, the result does not prove conclusively that the sum of the angles of the triangle is less than 180°; it might be taken to prove instead that the figure in question is simply not a triangle—and that we must adjust our views about light rays and measuring rods accordingly. Around the end of the nineteenth century, the French philosopher-scientist Henri Poincaré, following this line of thought, concluded that *any* apparent deviations from Euclidean geometry could always be explained away much as we have just indicated—that is, by saying that any disagreement with Euclidean spatial relations is a sure sign that something is wrong with our measuring rods or our experimental technique. In all such cases, we can account for the discrepancy from Euclidean geometry by means of some sort of perturbation affecting the measuring rods or other laboratory equipment.[7]

This is, perhaps, the plausible thing to say. It means that we can always, if we wish, adjust our experimental results to fit Euclidean geometry; we can, if we wish, preserve Euclidean geometry at all costs. Perhaps that is what should be done in such situations. Kant would certainly have agreed. As already mentioned, Kant maintained that Euclidean geometry is synthetic a priori—necessarily true—and that all of our experiences *must* fit the Euclidean framework. Accordingly, any kind of apparent conflict between physical measurement and Euclidean geometry would have to be resolved in favor of Euclidean geometry.

It is a rather common error to suppose that the discovery of non-Euclidean geometries in and of itself constitutes a refutation of Kant's doctrine of the synthetic a priori character of Euclidean geometry. Even in the light of the relative consistency proof of Euclidean and non-Euclidean geometries, the existence of alternative types of geometry does not refute Kant's thesis. Kant did not maintain that alternative geometries are self-contradictory, for that would make

Euclidean geometry analytic rather than synthetic.[8] And the fact that it is logically possible for our physical measurements to yield results in apparent conflict with Euclidean geometry does not necessarily rob Euclidean geometry of its a priori status; we have just seen how Euclidean geometry can be maintained in the face of any apparent empirical refutation. The fact that all of these different geometries are on a par with respect to logical consistency does not mean that they are equally suitable for the description of the physical world. Kant's thesis is simply a denial of the possibility of using any geometry other than Euclid's for the visualization of spatial relations.

This view is extremely tempting, and it is widely held today by people who may not even recognize it as Kantian. Kant maintained that Euclidean geometry is a necessary form of spatial intuition. Stated less pretentiously, this view runs somewhat as follows: "Non-Euclidean geometry may be an amusing game for mathematicians to play, but you cannot really conceive of a non-Euclidean space. You cannot imagine what it would be like for our three-dimensional space to be non-Euclidean. To be sure, we can picture a two-dimensional non-Euclidean surface, but that is because we can stand outside of it in our three-dimensional Euclidean space and observe the curvature of the surface as it is embedded in three-dimensional space. But we cannot picture the curvature of our whole three-dimensional space because we cannot imagine stepping off into four-dimensional space." This common sort of statement seems to me to express Kant's view that Euclidean geometry is *the* necessary form of spatial intuition. While all of the geometries are on an equal footing from a logical standpoint, they are not on an equal footing epistemologically. Non-Euclidean geometries cannot be visualized, Kant says, and so they cannot be used for the description of spatial relations. *Physical* space cannot be conceived to be non-Euclidean. Euclidean geometry, according to Kant, enjoys an insuperable epistemological advantage over the other types of geometry precisely because we cannot visualize a three-dimensional non-Euclidean space.

In order to deal with this Kantian doctrine we must first try to determine whether it is true that we cannot visualize non-Euclidean spaces. Even though it may be possible to describe physical space in Euclidean terms, is it necessary to do so? Can we organize our external experiences in Euclidean space alone, or is it also possible to do so in a non-Euclidean framework? If we are to come to terms with such questions, we must become quite clear on what we are to mean by "visualization."

There are no special difficulties in visualizing two-dimensional spaces, whether they be Euclidean or non-Euclidean. We can literally see the surfaces or we can call them up in our imagination. In either case, they are seen from without as two-dimensional manifolds embedded in three-dimensional space.[9] And the curvature of the non-Euclidean surfaces can readily be seen as a departure from the shape of the Euclidean plane. Let us call this *external visualization*.

It is easy to suppose that there is no particular problem in visualizing a Euclidean space of three dimensions, but clearly this is very different from the external visualization of a two-dimensional space. We cannot step outside of our three-dimensional space into a four-dimensional space in order to visualize externally *either* a three-dimensional Euclidean space *or* a three-dimensional non-Euclidean space. Instead, we must formulate an appropriate conception of *internal visualization* in order to understand what is involved in the visualization of a three-dimensional space of any variety. If we visualize three-dimensional Euclidean space, we do so from the standpoint of beings confined within that space; if we want to visualize three-dimensional non-Euclidean space, we must likewise do so from within. It was Hermann von Helmholtz, a nineteenth-century scientist and philosopher, who first saw this point and formulated a suitable concept of internal visualization. To visualize a space *internally*, he said, is simply *to imagine the kinds of experiences one would have if he were living in such a space*. We internally visualize a Euclidean space very easily because we are used to the kinds of experience people who are living in such a space have on a regular basis.[10] But obviously this capability in no way entails an ability to view our three-dimensional universe from a vantage point in a four-dimensional super-space.

With Helmoholtz's distinction between internal and external visualization clearly in mind, we can now see how it is possible to visualize three-dimensional non-Euclidean spaces. To make this point somewhat easier let us go back to the two-dimensional case. Imagine some two-dimensional creatures whose existence is confined entirely within a two-dimensional surface. Suppose that they can move around and make measurements (using two-dimensional instruments) in their two-dimensional space just as we can move about and make measurements in our three-dimensional space. They can no more leave their two-dimensional world to view it from a three-dimensional standpoint than can we leave our three-dimensional world to view it from a four-dimensional standpoint. Now, if it should happen that these two-dimensional beings inhabit the surface of a sphere they would be unable to see the curvature as we can from our external vantage-point, but they could detect the curvature indirectly. If they were to lay out triangles and circles, measuring angular sums and diameters and circumferences, they would find that the angular sum is always greater than two right angles, and that the ratio of circumference to diameter is always less than π. They would find, moreover, that the amount of the discrepancy depends upon the size of the figure. In such a case, they would have detected the curvature of the spherical surface without having seen it externally. For two-dimensional beings to visualize internally this type of non-Euclidean space *is* to imagine getting the foregoing sorts of results in making measurements. The internal visualization of Euclidean and Lobachevskian spaces of two dimensions is completely analogous.[11]

Similar considerations apply to the internal visualization of three-dimensional Euclidean and non-Euclidean spaces as well. We do not find it inconceivable that, in three-dimensional space, the measurement of the angles of triangles, regardless of their orientation in space, might result in a sum different from two right angles. Nor is it inconceivable that this result might vary with the size of the triangle. It is not inconceivable that measurements of circumferences and diameters of circles might yield a ratio other than π, whose value again depends upon the size of the figure. Analogous departures from Euclidean relations might also occur regarding the relation between the radius and the surface or the radius and volume of a sphere. These sorts of considerations show how, in principle, we could make a geometrical survey of our three-dimensional universe—in a manner quite analogous to our imaginary two-dimensional beings—in an effort to ascertain the geometrical structure of physical space.[12] As yet, the empirical results are inconclusive, but we can imagine what they would have to be like to indicate any of the types of geometry we have discussed. According to Helmholtz's conception, the ability to visualize internally spaces characterized by the different types of geometry consists in this very ability to imagine sets of experiences of the sort just described.[13]

It is extremely difficult, we must admit, to accept psychologically the import of Helmholtz's distinction between internal and external visualization, and to apply it to the problem of visualization of a three-dimensional non-Euclidean space. We find no difficulty in making the transition from *external* visualization of a two-dimensional Euclidean plane to the *internal* visualization of a three-dimensional Euclidean space, because all of our everyday spatial experiences since infancy have fit nicely into the Euclidean framework. If our universe has a non-Euclidean structure, it reveals itself only on a very large scale, and the deviation from Euclidean structure is too slight in moderate-sized regions to be noticed in ordinary experience. Because the transition from external to internal visualization is so easy in the Euclidean case, we may not even be aware of any distinction between the two types of visualization.

When, however, we turn to the problem of visualizing a non-Euclidean three-dimensional space, we may find the transition from the external visualization of the two-dimensional surface to the internal visualization of the three-dimensional space a severe psychological strain. We may feel, in fact, that there is something phony about translating the problem of "visualizing" or "imagining" a three-dimensional non-Euclidean space into the problem of internal visualization in Helmholtz's sense. It may seem simply impossible to imagine or picture a three-dimensional non-Euclidean space, no matter how successfully we can conceive intellectually the possibility of physical measurements that fit a non-Euclidean geometry. We must not forget, however, the psychological power of our lifelong conditioning to the Euclidean framework. If we had grown up in a world in which non-Euclidean relationships were a matter of daily experience, then it seems likely that the visualization of a three-dimensional non-Euclidean space would pose no more psychological difficulty than does the visualization of three-dimensional Euclidean space for us. If the experiences we must imagine (with some serious expenditure of intellectual effort) in order to visualize internally a three-dimensional non-Euclidean space were extremely familiar, this very same internal visualization would, it seems safe to say, be easy and psychologically satisfying. It is, therefore, a matter of psychological fact—not a priori necessity—that we have no problem internally visualizing a three-dimensional Euclidean space, while the visualization of a three-dimensional non-Euclidean space seems at first to present insuperable difficulties. Recognition of the distinction between internal and external visualization, and appreciation of the epistemological significance of that distinction, is a philosophical accomplishment of major proportions. Only after we have fully understood this point can we see that the problem of visualizing three-dimensional non-Euclidean space is, indeed, the problem of internal visualization in precisely the sense we have been discussing.

According to the foregoing results, we can visualize three-dimensional non-Euclidean spaces in precisely the same sense as we can visualize three-dimensional Euclidean spaces. Euclidean and non-Euclidean geometries are, therefore, on an equal footing from the standpoint of visualizability as well as from the purely formal standpoint. This conclusion, which goes far beyond mere logical parity, *does* constitute a serious blow to the Kantian doctrine that Euclidean geometry is synthetic a priori. Of course, it is always possible, as we have seen, to explain away any apparent non-Euclidean results by alleging the perturbation of measuring devices, but it is not necessary to do so, because we can visualize non-Euclidean spaces. To revert to Kantian terminology, non-Euclidean geometry thus emerges as a possible form of spatial intuition.

Although we have found that Euclidean and non-Euclidean geometries are formally and epistemologically admissible, there remains the question of which geometry correctly describes the physi-

THE TWIN SISTERS: PHILOSOPHY AND GEOMETRY

cal space of our universe. At most we have shown that none of these geometries can be ruled out a priori; we have not really shown how to answer the a posteriori problem of which one to adopt. The question is a vexing one because, as we have seen, empirical results of measurement do not seem to offer unambiguous answers. Any measurement which would seem to provide a non-Euclidean result can be interpreted as fitting into a Euclidean framework, and any measurement which seems to provide a Euclidean result can be adapted to a non-Euclidean framework. Whatever results we get by using light rays, solid rods, and similar instruments for making measurements can be explained away, for we can always say that they are perturbed. We can always claim that light rays bend and that solid rods shrink and expand. Poincaré has made this point very forcefully.

Consider another fanciful example, which was put forth by Hans Reichenbach. Imagine that world A, shown in cross section in Figure 7, is a two-dimensional world consisting of a flat plane with a hump in the middle. Again, imagine that this world is inhabited by two-dimensional creatures who move about and make measurements in an attempt to survey it geometrically. In the peripheral regions of their space, they would find that it has Euclidean characteristics—the ratio of circumference to diameter of a circle is always π, the angles of the triangle always add up to 180°, and so on. In the central part, where the hump is located, they would find that their space has the geometrical properties of the surface of a sphere. In the region where the hump joins smoothly with the flat plane, they would find characteristics rather like those of the saddle surface. Moving about in their space and making such measurements, they could find out that their space has precisely the kind of curvature just described, though they would be unable to form a three-dimensional mental image like ours.

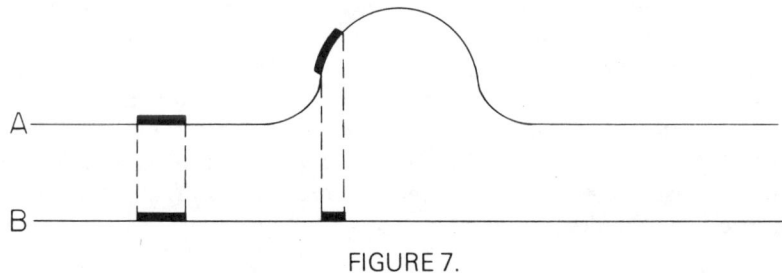

FIGURE 7.

Imagine, at the same time, beings who live in world B, located directly below world A, with measuring rods that behave in a peculiar fashion. Whenever these rods are moved from one place to another, they behave exactly as if they were vertical projections of measuring rods in world A above. Since *all* material bodies are assumed to expand and contract in the same manner, these beings would be quite unaware of such changes in their measuring rods. Obviously, the creatures in world B would get exactly the same results from their measurements as the inhabitants of world A. Now, this shows quite clearly that there are two interpretations available. A two-dimensional Poincaré living in world B might say, "Wait a minute—it is not necessarily true that our world is a flat surface with a hump in it. Maybe our world is one in which the measuring rods behave in a very odd manner, contracting and expanding in such a way that this world appears to have a hump when in fact it is perfectly flat." None of his countrymen in world B could refute him, as we saw in our discussion of Gauss's experiment. At the same time, precisely the same argument would be open to any inhabitant of world A as well, for the inhabitants of both worlds have precisely the same experiences. We see that, although these two worlds look very different from our Godlike external vantage point, from the standpoint of beings confined within these two worlds there is absolutely no way of distinguishing one from the other. Indeed, these are identical worlds! We have merely offered two equivalent ways of describing the same spatial facts. Henceforth we shall refer, not to two worlds, but rather to two descriptions, A and B, of the world of Figure 7.

It might be objected, however, that we have not yet come to grips with the basic problem—namely, the question of whether the measuring rods really do expand and contract as they are moved about, or whether they remain the same length. If we knew the answer to that question we would know whether the world really had a hump or not. But how can we find out? We do know that measuring rods expand and contract under the influence of certain kinds of forces, for example, as a result of changes of temperature. This kind of expansion and contraction is empirically detectable because it affects different materials in different ways. A copper rod, for example, expands a different amount and at a different rate from a glass rod. Forces whose influence depends upon the chemical constitution of the object affected are known as *differential forces;* their results are called *differential effects.* Whenever precise measurement is attempted, great care must be taken to correct for differential effects upon the instruments employed.

When we raise a question, however, about the expansion or contraction of measuring rods which are moved from place to place, we are assuming that such rods are free from the influence of any differential forces or that suitable corrections have been made for any differential effects. If these rods are subject to such expansion and contraction it is a *universal effect* which depends only upon the position of the object. It is due to a *universal force* that affects all substances in the same manner and to the same degree.[14] The fact that such effects are universal, not occurring differentially in different objects, guarantees that they will be empirically undetectable. To the question of whether two measuring rods located at different places are really the same length, or whether the same measuring rod keeps the same length as it is moved from place to place, there is no possible empirical answer. Consequently, Reichenbach argues, the question is not a factual question at all; it must be construed as a demand for a definition or convention. The concept which requires definition is *congruence;* we must define equality of spatial intervals when these intervals are located at different places. One natural definition of congruence results when we say that a solid rod remains congruent to itself—that is, it retains the same size—wherever it is located and however it is oriented in space (correcting, of course, for any differential effects). This is not, however, the only possible definition of congruence. Our two-dimensional Poincaré got rid of the hump by introducing a different definition of congruence.

Reichenbach's analysis shows, I think, that congruence cannot be discovered, but must be defined, *if* equality or inequality of length (or distance) is constituted by, and has no meaning apart from, the behavior of solid rods and other physical measuring devices. This means, in effect, that length is not a characteristic of space, but of the bodies that occupy space. Such a view differs fundamentally from Newton's account. According to Newton, absolute space is some sort of entity that is entirely distinct from, and not to be confused with, material bodies that may be used to measure space. Just as a table should not be identified with the ruler used to measure it, so also must absolute space be regarded as something separate from our measuring rods. Absolute space is taken by Newton to be a container for material objects, in no way depending upon them for its nature or existence. Hence, in a Newtonian framework, it is still appropriate to ask whether a measuring rod really expands or contracts when moved from place to place, since it is meaningful on the Newtonian view to ask whether it occupies the same amount of absolute space in one place as in the other. At one time, the two ends of the measuring rod are located at points P_1 and P_2; at a later time, the two ends of the same measuring rod are located at P_3 and P_4. The Newtonian question is this: is the quantity of absolute space in the interval P_1P_2 equal to the quantity of absolute space in the interval P_3P_4? According to Reichenbach's analysis, we are free to *define* the two as equal on the basis of the behavior of the measuring rods; according to Newton's approach we must *discover* whether the two amounts of space are equal. This issue resolves itself into the question of whether physical space has an intrinsic metrical structure. If it does, there is a true answer to the question of whether measuring rods change their size when they are transported to different locations. If space has such an intrinsic structure, then it also has a determinate intrinsic geometry. If no such intrinsic structure exists, it is open to us to define congruence, and different definitions of congruence may yield different geometries as correct descriptions of the same space, as we saw in connection with the world with a hump.

It was Riemann who first clearly posed the question of whether space has an intrinsic metric. His answer—the correct answer, I believe—was negative. Although his reasoning was somewhat faulty, due to the fact that Georg Cantor's set theory had not been developed at the time, it seems basically to be on the right track. I shall not elaborate in detail in this chapter for this problem is closely related to one of Zeno's paradoxes (not one of the famous paradoxes of motion, but a less familiar paradox of plurality), which we shall discuss at some length in the next chapter. But the general approach is this. A continuous finite line segment, regardless of length, contains an infinite number of points; indeed, it can be shown that any finite line segment contains precisely the same infinite number of points as any other. These points are, moreover, ordered in precisely the same way in any two such segments. This shows that the one line segment is *isomorphic* to the other (regardless of length)—that is, they have the same structure. Since all finite line segments are identical in internal structure, difference of length must somehow be imposed on an extrinsic basis. This is precisely where Reichenbach's definition of congruence comes in; it provides a basis for comparison of length by means of solid bodies which function as measuring rods.[15]

It turns out, then, that the question of whether the measuring rod really changes its length when moved from place to place is without significance.[16] Hence, the question of whether space is really Euclidean or non-Euclidean loses its significance as well. But this does not mean that whatever we might say about the geometrical structure of physical space is without significance. Nor does it mean that we are free to settle all questions about the geometry of space by means of definitions. Rather, the question of what geometry correctly describes physical space is an incomplete question unless a definition of congruence is provided. In the world of Figure 7, we may adopt the definition of congruence which says that the measuring rod does not change in length as it is moved about, and then determine by empirical measurements that the geometry is non-Euclidean. Or, we may adopt the definition of congruence which says that the measuring rods behave as stipulated by our two-dimensional Poincaré, and ascertain by empirical measurement that the geometry is Euclidean. These two different definitions of congruence yield two different descriptions of the world. To the question, "Which of these descriptions is correct?" the answer is that both are. They are *equivalent descriptions.*[17] Every observable fact which tends to support the one also tends to support the other; any observable fact that would tend to refute the one would also tend to refute the other. The situation is somewhat analogous (though not completely so) to the question, "Is New York really to the right of Chicago?" This question obviously has no definite answer until we specify a vantage point. To a vacationer returning to Cleveland from Florida, the answer is affirmative; to an explorer headed toward the same city on his return from the North Pole, the answer is negative. But no one finds any difficulty in seeing why either a *yes* or a *no* can be correct, depending upon the way the question is specified.

It is equally obvious that the incompleteness of such questions does not mean that every conceivable answer is correct. It is incorrect to say that Chicago is to the left of New York from the standpoint of our returning explorer. It is equally incorrect to say that the geometry of the world of Figure 7 is Euclidean given that the measuring rods do not shrink or expand as they are moved about. This would be a correct description of some other world, but it is a false description of the world of Figure 7; such a world is pictured in Figure 8. This same world can correctly be described as non-Euclidean (having a hump) if congruence is defined in terms of the vertical projection shown in description B.

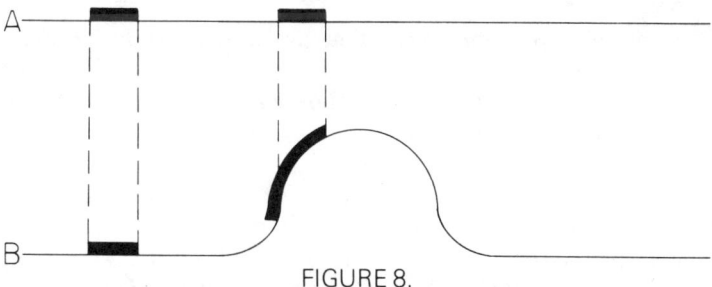

FIGURE 8.

Given the fact that we have a choice between descriptions A and B of Figure 7, on what basis should we choose? Truth or falsity is not a ground for choice, since both are true descriptions of the world of Figure 7 (and both would be false descriptions of the world of Figure 8). The choice must be made on such grounds as simplicity, economy, and elegance. The outcome of the choice may be open to dispute. Some people might say that a definition of congruence which involves no "universal forces" is simpler than one which has them; hence, description A is preferable. Others might say that Euclidean geometry is simpler than non-Euclidean geometry; hence, description B is preferable. Although Poincaré seems to have chosen the second alternative, Einstein (in the general theory of relativity) has chosen the first, and a majority of contemporary relativity physicists seem to agree. But the fundamental philosophical point is this: a description of physical space involves at least two elements, a specification of congruence relations and a geometry. When congruence has been defined, the question of which geometry applies is a determinate empirical question that can be settled in principle by physical measurement. One may instead choose a geometry by definition—we have discussed ways of preserving Euclidean geometry come what may. Such a stipulation, though not uniquely singling out a congruence relation, does severely limit the admissible candidates. Which ones are eligible must be ascertained by the empirical facts. Roughly speaking, the geometrical description of physical space involves two factors, a specification of congruence and a type of geometry, one of which can be chosen as a matter of convention. The remaining ingredient involves the ascertainment of empirical facts.

We may summarize the situation in the following way. There are various mathematical systems—systems of geometry—Euclidean and non-Euclidean. We want to try to apply these systems to the space in which we live. In order to do this we must say what we mean by the various terms that occur in these systems—"straight line," "intersection," "point," and most especially, "congruence."

If we are to make measurements, we must have measuring instruments of some sort. Obviously, we have to define a unit of measurement; we have to say whether we are going to measure in meters, centimeters, feet, yards, and so forth. Furthermore, having once defined the unit of measurement, we have to say what it means to talk about *equality of length*—that is to say, *congruence*—at different places. And that is nothing more than the question of what happens to the measuring rods as they are moved from place to place. Do they stay the same size? Do they change? This is not a question of fact; it is a question of definition.[18] If we do not say what we mean by sameness of size, then there is no answer to the question of whether the world is Euclidean or non-Euclidean. We can say either because we are free to make different definitions regarding length and congruence. If, however, we pin down the concept of congruence by specifying physically what we mean by the length of a measuring rod as it is moved from place to place, then the question as to the geometry of our space becomes a straightforward, though difficult, empirical one. The function of definitions—often called *coordinating definitions*—is to provide a link between the abstract mathematical concepts that appear in the geometrical systems and the physical objects that populate the physical world. Without such coordinating definitions, there is no connection whatever between the mathematical system and the real world.

What are the philosophical conclusions to be drawn from this long story—a story beginning in 600 B.C. and tracing the relationships between philosophy and geometry for some 2500 years? Is the philosophical dream of Plato a realizable one? Can we, after all these centuries, come to the same conclusion as Plato did—that geometry provides us real knowledge of the physical world based solely upon reason? Is geometrical knowledge both synthetic and a priori as Kant claimed?

The answer to all these questions is, I think, negative. Although the discovery of non-Euclidean geometries does not, by itself, refute Kant's thesis, the existence of several alternative geometrical systems does force upon us a distinction between two ways of looking at geometry. In the first place, we can view a geometry as a mathematical system in which we lay down our postulates, whether they be the postulates of Euclidean or non-Euclidean geometry, and then make our deductions from them. From this standpoint we are not in the least concerned with whether the postulates are true; the only question is the logical relation between the postulates and the theorems that follow from them. Indeed, within this type of abstract mathematical system, the postulates and theorems are neither true nor false, for they contain primitive terms such as "point," "straight line," and "congruent" which, strictly speaking, have no meaning at all prior to the introduction of coordinating definitions which assign meanings to them. A system of this kind, containing uninterpreted primitive terms, is a system of *pure geometry;* as such, it is entirely a priori. This does not mean that the postulates (or theorems) are a priori truths. The *logical relation* between the postulates and the theorems is the a priori aspect of pure geometry. The proof of the theorems does not depend in any way upon experiment or experience. They are proved by deduction from the postulates. Such abstract systems of pure mathematics are not, however, synthetic; by themselves they convey no information whatever about the physical world. In the absence of coordinating definitions of the geometrical terms, they bear no relation to the physical world.

In the second place, we can look at geometry as a mathematical system to be used for the purpose of describing the world. When we ask what geometry describes the physical world, we are looking at geometry as *applied geometry*. And now, as we have seen, *if we make some appropriate coordinating definitions to begin with*, there is a definite answer to the question, "What geometry describes physical space?"—a question that may be hard to answer, but which is, nevertheless, a perfectly reasonable physical question. It becomes a reasonable question because our coordinating definitions associate such previously undefined geometrical concepts as straight line and congruence with such physical objects as light rays and measuring rods. This type of geometry—applied geometry—does give us knowledge of the physical world. It does inform us about the spatial structure of our universe. It is synthetic, but unfortunately, it is not a priori. Pure reason, unaided by the senses, cannot tell us how light rays and solid rods will behave; there is no way to tell a priori which geometry is the actual geometry of physical space. That can only be decided by empirical observations and physical experiments.

We find, consequently, that there is a type of geometry which is a priori—namely, pure geometry, an axiomatic system of pure mathematics. There is a type of geometry that is synthetic—namely, applied geometry, a system used to describe the physical world. Unfortunately, pure geometry is not applied geometry, and applied geometry is not pure geometry; there is no type of geometry which is both a priori and synthetic. The clear recognition of this fundamental logical distinction between pure and applied mathematics came about as a result of the development of non-Euclidean geometries. And a full appreciation of its significance requires a careful analysis of the nature and function of coordinating definitions.

Thus, the hope that geometry held out to philosophers for well over two millenia—the hope that pure reason could give us basic information about the nature of the world we live in—is seen ultimately to have been dashed, and by a development in the field of geometry itself. To know what kind of world we live in, to achieve scientific knowledge of it, we must look to the empirical sciences. Pure mathematics, by itself, will never yield such results. Mathematics has to be applied within the natural sciences, and in this context it turns out to be an enormously useful tool for our understanding of reality. But without the observation and experiment which form the indispensable core of the empirical sciences, pure mathematics can tell us nothing about the world we live in.

SUGGESTED READINGS

1. Adler, Irving. *A New Look at Geometry.* New York: The New American Library (Signet Y3225), 1966.
2. Aleksandrov, A.D., Kolmogorov, A.N., and Lavrent'ev, M.A., eds. *Mathematics: Its Content, Methods, and Meaning.* Cambridge, Mass.: The M.I.T. Press, 1963. Especially Chap. XVII, "Non-Euclidean Geometry," by A.D. Aleksandrov.
3. Bonola, Roberto. *Non-Euclidean Geometry.* New York: Dover Publications, Inc., 1955. Contains also Bolyai, "The Science of Absolute Space," and Lobachevski, "The Theory of Parallels."
4. Euclid. *The Thirteen Books of Euclid's Elements.* Sir Thomas L. Heath, trans. New York: Dover Publications, Inc., 1956. Contains extensive commentary on Euclid.
5. Grünbaum, Adolf. *Philosophic Problems of Space and Time.* New York: Alfred A. Knopf, 1963. 2nd ed., Boston: D. Reidel Publishing Co., 1974.
6. Hilbert, David. *Foundations of Geometry,* 2nd ed. La Salle, Ill.: Open Court, 1971.
7. Poincaré, Henri. *Science and Hypothesis.* New York: Dover Publications, Inc., 1952.
8. Reichenbach, Hans. *The Philosophy of Space and Time.* New York: Dover Publications. Inc., 1958.
9. Smart, J.J.C. *Problems of Space and Time.* New York: The Macmillan Co., 1964. An anthology of philosophical readings.
10. van Fraassen, Bas C. *An Introduction to the Philosophy of Time and Space.* New York: Random House, 1970.
11. Wolfe, Harold E. *Introduction to Non-Euclidean Geometry.* New York: Holt, Rinehart and Winston, Inc., 1945.
12. Young, J.W.A., ed. *Monographs on Topics of Modern Mathematics.* New York: Dover Publications, Inc., 1955. Article I, "The Foundations of Geometry" by Oswald Veblen provides a modern axiomatization of Euclidean geometry, and Article III, "Non-Euclidean Geometry" by Frederick S. Woods, presents non-Euclidean geometry axiomatically by suitable modifications of Veblen's axioms for Euclidean geometry.

Chapter Two

A CONTEMPORARY LOOK AT ZENO'S PARADOXES*

The intellectual heritage bequeathed to us by the ancient Greeks was rich indeed. The science of geometry and the entire course of Western philosophy, as we have noted, both had their beginnings with Thales. Both enjoyed fantastic development at the hands of his early successors, achieving a surprising degree of perfection during antiquity. During the same period, Aristotle provided the first systematic development of formal logic. But the fertile soil from which all of this grew also gave rise to a series of puzzles which have challenged successive generations of philosophers and scientists right down to the present. These are the famous paradoxes of Zeno of Elea who flourished about 500 B.C.[1]

Zeno was a devoted disciple of the philosopher Parmenides, who had held that reality consisted of one undifferentiated, unchanging motionless whole which was devoid of any parts. Motion, change, and plurality were, according to him, mere illusions. Not too many philosophers could accept this view, and Parmenides was apparently the object of some ridicule from those who disagreed. Zeno's main purpose, it is reported, was to refute those who made fun of his master. His aim was to show that those who believed in motion, change, and plurality were involved in even greater absurdities. Out of perhaps forty such puzzles that he propounded, fewer than ten have come down to us, but they involve some very subtle difficulties. Since motion involves the occupation of different *places* at different *times*, these paradoxes strike at the heart of our concepts of space and time.

Bertrand Russell once remarked that "Zeno's arguments, in some form, have afforded grounds for almost all theories of space and

---

*Much of the material in this chapter has been adapted from my Introduction in Wesley C. Salmon, ed., *Zeno's Paradoxes*, copyright © 1970, The Bobbs-Merrill Co., Inc. By permission of the publisher.

time and infinity which have been constructed from his time to our own."[2] This statement was made in 1914, in an essay which contains a penetrating analysis of the paradoxes, but as we shall see, there were problems inherent in these puzzles that escaped even Russell. Such difficulties, in fact, have a direct bearing upon our foregoing discussions of space and geometry, revealing deep problems that we have barely mentioned. Moreover, before this chapter is finished, I shall be reporting a new Zeno-type puzzle which first came to light during the present decade!

The following paradoxes fall into two main categories, paradoxes of *motion* and paradoxes of *plurality*. The paradoxes of motion are the more famous ones, and I shall begin with them.

THE PARADOXES OF MOTION

Our knowledge of the paradoxes of motion comes from Aristotle who, in the course of his discussions, offers a paraphrase of each. Zeno's original formulations have not survived.[3]

1. *Achilles and the Tortoise.* Imagine that Achilles, the fleetest of Greek warriors, is to run a footrace against a tortoise. It is only fair to give the tortoise a head start. Under these circumstances, Zeno argues, Achilles can never catch up with the tortoise, no matter how fast he runs. In order to overtake the tortoise, Achilles must run from his starting point A to the tortoise's original starting point T_0 (see Figure 1). While he is doing that, the tortoise will have moved ahead to T_1. Now Achilles must reach the point T_1. While Achilles is covering this new distance, the tortoise moves still farther to T_2.

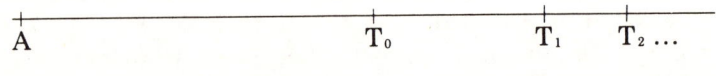

A T_0 T_1 T_2 ...

FIGURE 1.

Again, Achilles must reach this new position of the tortoise. And so it continues; whenever Achilles arrives at a point where the tortoise *was*, the tortoise has already moved a bit ahead. Achilles can narrow the gap, but he can never actually catch up with him. This is the most famous of all of Zeno's paradoxes. It is sometimes known simply as "The Achilles."

2. *The Dichotomy.* This paradox comes in two forms, progressive and regressive. According to the first, Achilles

cannot get to the end of any racecourse, tortoise or no tortoise; indeed, he cannot even reach the original starting point T_0 of the tortoise in the previous paradox. Zeno argues as follows. Before the runner can cover the whole distance he must cover the first half of it (see Figure 2).

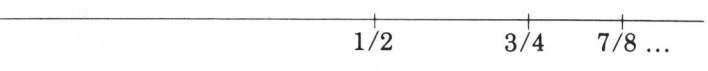

FIGURE 2.

Then he must cover the first half of the remaining distance, and so on. In other words, he must first run one-half, then an additional one-fourth, then an additional one-eighth, etc., always remaining somewhere short of his goal. Hence, Zeno concludes, he can never reach it. This is the progressive form of the paradox, and it has very nearly the same force as Achilles and the Tortoise, the only difference being that in the Dichotomy the goal is stationary, while in Achilles and the Tortoise it moves, but at a speed much less than that of Achilles.

The regressive form of the Dichotomy attempts to show, worse yet, that the runner cannot even get started. Before he can complete the full distance, he must run half of it (see Figure 3). But before he can complete the first half, he must run half of that, namely, the first quarter.

$$\ldots 1/8 \quad 1/4 \quad 1/2$$

FIGURE 3.

Before he can complete the first quarter, he must run the first eighth. And so on. In order to cover any distance no matter how short, Zeno concludes, the runner must already have completed an infinite number of runs. Since the sequence of runs he must already have completed has the form of a regression,

$$\ldots, \frac{1}{16}, \frac{1}{8}, \frac{1}{4}, \frac{1}{2},$$

it has no first member, and hence, the runner cannot even get started.

3. *The Arrow.* In this paradox, Zeno argues that an arrow in flight is always at rest. At any given instant, he claims, the arrow is where it is, occupying a portion of space equal

to itself. During the instant it cannot move, for that would require the instant to have parts, and an instant is by definition a minimal and indivisible element of time. If the arrow did move during the instant it would have to be in one place at one part of the instant, and in a different place at another part of the instant. Moreover, for the arrow to move during the instant would require that during the instant it must occupy a space larger than itself, for otherwise it has no room to move. As Russell says, "It is never moving, but in some miraculous way the change of position has to occur *between* the instants, that is to say, not at any time whatever."[4] This paradox is more difficult to understand than Achilles and the Tortoise or either form of the Dichotomy, but another remark by Russell is apt: "The more the difficulty is meditated, the more real it becomes."

4. *The Stadium.* Consider three rows of objects A, B, and C, arranged as in the first position of Figure 4. Then, while row A remains at rest, imagine rows B and C moving in opposite directions until all three rows are lined up as shown in the second position. In the process, C_1 passes twice as many B's as A's; it lines up with the first A to its left, but with the second B to its left. According to Aristotle, Zeno concluded that "double the time is equal to half."

	First Position				*Second Position*		
	A_1	A_2	A_3		A_1	A_2	A_3
B_1	B_2	B_3			B_1	B_2	B_3
	C_1	C_2	C_3		C_1	C_2	C_3

FIGURE 4.

Some such conclusion would be warranted if we assume that the time it takes for a C to pass to the next B is the same as the time it takes to pass to the next A, but this assumption seems patently false. It appears that Zeno had no appreciation of relative speed, assuming that the speed of C relative to B is the same as the speed of C relative to A. If that were the only foundation for the paradox we would have no reason to be interested in it, except perhaps as a historical curiosity. It turns out, however, that there is an interpretation of this paradox which gives it serious import.

Suppose, as people occasionally do, that space and

time are atomistic in character, being composed of space-atoms and time-atoms of non-zero size, rather than being composed of points and instants whose size is zero.[5] Under these circumstances, motion would consist in taking up different discrete locations at different discrete instants. Now, if we suppose that the A's are not moving, but the B's move to the right at the rate of one place per instant while the C's move to the left at the same speed, some of the C's get past some of the B's without ever passing them. C_1 begins at the right of B_2 and it ends up at the left of B_2, but there is no instance at which it lines up with B_2; consequently, there is no time at which they pass each other—it never happens.

It has been suggested that Zeno's arguments fit into an overall pattern.[6] Achilles and the Tortoise and the Dichotomy are designed to refute the doctrine that space and time are continuous, while the Arrow and the Stadium are intended to refute the view that space and time have an atomic structure. The paradox of plurality, which will be discussed later, also fits into the total schema. Thus, it has been argued, Zeno tries to cut off all possible avenues to escape from the conclusion that space, time, and motion are not real but illusory.

It is extremely tempting to suppose, at first glance, that the first three of these paradoxes at least arise from understandable confusions on Zeno's part about concepts of the infinitesimal calculus. It was in this spirit that the American philosopher C. S. Peirce, writing early in the twentieth century, said of Achilles that "this ridiculous little catch presents no difficulty at all to a mind adequately trained in mathematics and logic."[7] There is no reason to think he regarded any of Zeno's other paradoxes more highly.

We should begin by noting that, although the calculus was developed in the seventeenth century, its foundations were beset with very serious logical difficulties until the nineteenth century—when Cauchy clarified such fundamental concepts as functions, limits, convergence of sequences and series, the derivative, and the integral; and when his successors Dedekind, Weierstrass, et al., provided a satisfactory analysis of the real number system and its connections with the calculus. I am firmly convinced that Zeno's various paradoxes constituted insuperable difficulties for the calculus in its pre-nineteenth-century form, but that the nineteenth-century achievements regarding the foundations of the calculus provide means which go far toward the resolution of Zeno's paradoxes. Let us see what light these purified concepts can throw on the paradoxes of motion.[8]

The sum of an infinite series. It is hard to guess how deep or subtle Zeno's actual reasoning was; experts differ on the point.[9] It may have been that Zeno's original version of Achilles and the Tortoise involved the following sort of argument: since Achilles must traverse an infinite number of distances, each greater than zero, in order to catch up with the tortoise, he can never do so, for such a process would take an infinite amount of time. Against this form of the argument Aristotle quite appropriately pointed out that the time span during which Achilles chases after the tortoise can likewise be subdivided into infinitely many non-zero intervals, so Achilles has infinitely many non-zero time intervals in which to traverse the infinitely many non-zero space intervals. But this response can hardly be adequate, for the question still remains: how can infinitely many positive intervals of time *or* space add up to anything less than infinity? The answer to this question was not provided until Cauchy offered a satisfactory treatment of convergent series in the first half of the nineteenth century.

The first concept we need is the *limit* of an infinite sequence. An infinite sequence is simply an ordered set of terms $\{S_n\}$ which correspond in a one-to-one fashion with the positive integers—each term of the sequence being coordinated by the subscript n to a positive integer. The sequence is said to be *convergent* if it has a limit. To say that such a sequence has a limit means that there is some number L (the limit) such that the terms of the sequence become and remain arbitrarily close to that value as we run through the successive terms. More precisely, for any number ϵ greater than 0, there is some positive integer N such that for every term S_n with $n > N$, the difference between S_n and L is less than ϵ. In the sequence

$$\frac{1}{2}, \frac{1}{4}, \frac{1}{8}, \ldots, \frac{1}{2^n}, \ldots$$

the limit is 0, since the difference between the terms of the sequence and 0 is arbitrarily small for sufficiently large values of n. If, for example, we choose $\epsilon = 1/10$, by the time we reach the fourth term $S_4 = 1/16$, the difference between that term and L $(= 0)$ is less than $1/10$, and the difference remains less than $1/10$ for every subsequent member of the sequence. For $\epsilon = 1/100$, $|S_n - 0|$ is less than ϵ for $n = 7$, and the difference remains less than $1/100$ for every subsequent term. Similarly, ϵ may be chosen as small as we like, say $1/1,000,000$ or $1/1,000,000,000$, provided it is greater than zero, and there is some point in this sequence beyond which all remaining terms differ from L by less than ϵ. It is easy to show, by completely parallel reasoning, that the sequence

$$\frac{1}{2}, \frac{3}{4}, \frac{7}{8}, \ldots, 1 - \frac{1}{2^n}, \ldots$$

converges to the limiting value of 1.

A CONTEMPORARY LOOK AT ZENO'S PARADOXES

After the concept of the limit of a sequence has been defined, it can be used to define the sum of an infinite *series*. An infinite series is simply an infinite sequence of terms which are related to one another by addition; for example,

$$\frac{1}{2} + \frac{1}{4} + \frac{1}{8} + \ldots + \frac{1}{2^n}, + \ldots$$

Such a sum is not defined in elementary arithmetic, for ordinary addition is restricted to sums of finite numbers of terms, but this operation can be extended very naturally to an infinite series. In order to define the sum of an infinite *series*

$$s_1 + s_2 + s_3 + \ldots$$

we form the *sequence* of partial sums,

$$S_1 = s_1$$
$$S_2 = s_1 + s_2$$
$$S_3 = s_1 + s_2 + s_3$$
etc.

Each of these partial sums is a sum with a finite number of terms, and it involves only the familiar operation of addition from elementary arithmetic. We have already defined the limit of an infinite sequence. If the *sequence* of partial sums,

$$S_1, S_2, S_3, \ldots$$

has a limit, we say that the infinite *series*

$$s_1 + s_2 + s_3 + \ldots$$

is convergent, and we define its sum as the limit of the sequence of partial sums. This amounts to saying, intuitively, that the sum of a convergent infinite series is a number that can be approximated arbitrarily closely by adding up a sufficient (finite!) number of terms. Given this definition of the sum of an infinite series, it becomes perfectly meaningful to say that the infinitely many terms of a convergent series have a finite sum.

Both the first form of the Dichotomy and the Achilles paradoxes present us with infinite series to be summed. In the Dichotomy, for instance, it is shown that the runner, to cover a racecourse that is one mile in length, must cover the following series of non-overlapping distances:

$$\frac{1}{2} + \frac{1}{4} + \frac{1}{8} + \ldots$$

Each term of this series is greater than zero. We form the sequence of partial sums

$S_1 = \frac{1}{2}$

$S_2 = \frac{1}{2} + \frac{1}{4} = \frac{3}{4}$

$S_3 = \frac{1}{2} + \frac{1}{4} + \frac{1}{8} = \frac{7}{8}$

etc.

As we noted above, this sequence converges to the limit 1; that is the sum of this convergent infinite series. Achilles and the Tortoise is quite analogous. If Achilles can run twice as fast as the tortoise, and the tortoise has a head start of one-half of the course, the infinite series generated by Achilles running to each subsequent starting point of the tortoise is precisely the one we have just summed. To whatever extent these paradoxes raised problems about the intelligibility of adding up infinitely many positive terms, the nineteenth-century theory of convergent sequences and series resolved the problem.

Instantaneous velocity. An intial reaction to the paradox of the Arrow might be the suspicion that it hinges on a confusion between the concepts of instantaneous motion and instantaneous rest. Perhaps Zeno did feel that the only way for an arrow to be at a particular place was to be at rest—that the notion of instantaneous non-zero velocity was illegitimate. If Zeno argued—we have no way of knowing whether he did or not—that at every moment of its flight the arrow is at some place in its trajectory, and hence at every moment of its flight it has velocity zero, then he would have been correct in concluding that its velocity during the whole course of its flight would be zero, rendering the arrow motionless. Nineteenth-century mathematics showed, however, that one of these assumptions is incorrect. It is entirely intelligible to attribute non-zero instantaneous velocities to moving objects when an instantaneous velocity is understood as a derivative—namely, the rate of change of position with respect to time. This derivative is defined as the limit of the average velocity during decreasing non-zero intervals of time. Suppose, for example, that the arrow flies at a uniform speed. We find that in one second it covers ten feet, in one-tenth of a second it covers one foot, in one-hundredth of a second it covers one-tenth of a foot, and so on. As we take these *average* velocities over decreasing finite time intervals which converge to an instant t_1, the average velocities approach a limit of ten feet per second, and this is, by definition, the instantaneous velocity of the arrow at t_1. The same can be said for every moment during its flight; it travels its whole course at ten feet per second, and its velocity at each moment is ten feet per second. If Zeno felt that the only intelligible instantaneous velocity is zero, nineteenth-century mathematics proved him wrong.

The infinitesimal calculus was, of course, developed in the seventeenth century, and it made use of instantaneous velocities. These were, unfortunately, considered to be infinitesimal distances covered in infinitesimal times. It was against such notions that Berkeley leveled his broadside in *The Analyst*,[10] characterizing infinitesimals as "ghosts of recently departed quantities." It is possible that Zeno's Arrow paradox was also directed against just such a conception. If we try to conceive of finite motion over a finite distance during a finite time as being composed of a large number of motions over infinitesimal distances during infinitesimal times, enormous confusion is likely to ensue. How much space does an arrow occupy during an infinitesimal time? Is it just as large as the arrow, or is it a wee big larger? If it is larger, then how does the arrow get from one part of that space to another? And if not, then how can the arrow be moving at all? And how long is an infinitesimal time span? Does it have parts or not? If so, how can we characterize motion during its parts? If not, how can motion occur during this infinitesimal time? These are questions that Zeno and his fellow Greeks could not answer, and to which modern calculus prior to Cauchy had no satisfactory answer either. This is why I remarked earlier that nineteenth-century — not seventeenth-century — mathematics held an important key, in the concept of the derivative, to the resolution of Zeno's Arrow paradox.

Mathematical functions. There is, however, still an underlying problem about instantaneous velocity. We have seen how such a concept can be defined intelligibly, but this definition makes essential reference to what is happening at neighboring instants. Instantaneous velocity is defined as a limit of a sequence of average velocities over finite time intervals; without some information about what happens in these intervals we can say nothing about the instantaneous velocity. If we know simply that the center of the arrow was at the point s_1 at time t_1 we can draw no conclusion whatever about its velocity at that instant. Unless we know what the arrow was doing at other times close to t_1 we *cannot* distinguish instantaneous motion from instantaneous rest. It was just this consideration, I believe, which led the philosopher Henri Bergson to say that Zeno's Arrow paradox calls attention to the absurd proposition ". . . that movement is made of immobilities."[11] Bergson concluded that the Arrow paradox proves that the standard mathematical characterization of motion must be wrong. We must look at this argument a little more closely.

In modern physics, motion is treated as a functional relationship between points of space and instants of time. The formula for the motion of a freely falling body, for example, is

$$x = f(t) = \frac{1}{2}gt^2.$$

Such formulas make it possible, by employing the function f, to compute the position x given a value of time t. But to understand this treatment of motion fully, it is necessary to have a clear conception of mathematical functions. Before the nineteenth century there was no satisfactory treatment of functions; functions were widely regarded as things which moved or flowed. Such a conception is of no help in attempting to resolve Zeno's paradoxes; on the contrary, Zeno's paradoxes of motion constitute severe difficulties for any such notion of mathematical functions. The situation was dramatically improved when Cauchy defined a function as simply a pairing of numbers from one set with numbers from another set. The numbers of the first set are the *values of the argument*, sometimes called the *independent variable*; the numbers of the second set (which need not be a different set) are the *values of the function*, sometimes called the *dependent variable*. For example, the function $F(x) = y = x^2$ pairs real numbers with non-negative real numbers. With the number 2 it associates the number 4, with the number -1 it associates the number 1, with the number 1/2 it associates the number 1/4, and so forth. Now according to Cauchy, the mathematical function F simply *is* the set of all such pairs of numbers, namely,

x	$F(x) = x^2$
1	1
2	4
3	9
$\frac{1}{2}$	$\frac{1}{4}$
$\frac{1}{3}$	$\frac{1}{9}$
-2	4
-1	1
etc.	etc.

Similarly, the function f used to describe the motion of a falling body is nothing more or less than a pairing of the values of the position variable x with values of the time variable t. At $t = 0$, $x = 0$; at $t = 1$, $x = 16$; at $t = 2$, $x = 64$. This is how we say, in mathematical language, that a body starting from rest in the vicinity of the surface of the earth and falling freely travels 16 feet in the first second, 48 feet in the next second, and so on.

Let us now apply this conception of a mathematical function to the motion of an arrow; to keep the arithmetic simple, let it travel at the uniform speed of ten feet per second in a straight line, starting from $x = 0$ at $t = 0$. At any subsequent time t, its position $x = 10t$. Accordingly, part of what we mean by saying that the arrow moved from point A ($x = 10$) to point B ($x = 30$) is simply that it was *at A* when $t = 1$, and it was *at B* when $t = 3$. When we ask how it got from A to B, the answer is that it occupied each of the intervening points x ($10 < x < 30$) at suitable times t ($1 < t < 3$) — that is, satisfying the equation $x = 10t$. For example, when $t = 2$, the arrow was at the point C ($x = 20$). When we ask how it got from A to C, the answer is again: by occupying the intervening positions at suitable times. Notice that this answer is *not:* by zipping through the intervening points at ten feet per second. The requirement is that the arrow be *at* the appropriate point *at* the appropriate time — nothing is said about the instantaneous velocity of the arrow as it occupies each of these points. This approach has been appropriately dubbed "the at-at theory of motion." Once the motion has been described by a mathematical function that associates positions with times, it is then possible to differentiate the function and find its derivative, which in turn provides the instantaneous velocities for each moment of travel. But the motion itself is described by the pairing of positions with times alone. Thus, Russell was led to remark, "Weierstrass, by strictly banishing all infinitesimals, has at last shown that we live in an unchanging world, and that the arrow, at every moment of its flight, is truly at rest. The only point where Zeno probably erred was in inferring (if he did infer) that, because there is no change, therefore the world must be in the same state at one time as at another. This consequence by no means follows. . . ."[12]

What Russell is saying is basically sound, although he does perhaps phrase it overdramatically. It is not that the arrow is "truly at rest" during its flight; rather, the motion consists in being *at* a particular point *at* a particular time, and regarding each individual position at each particular moment, there is no distinction between being at rest at the point and being in motion at the point. The distinction between rest and motion arises only when we consider the positions of the body at a number of different moments. This means that, aside from *being at* the appropriate places at the appropriate times, there is no *additional* process of *moving* from one to another. In this sense, there is no absurdity at all in supposing motion to be composed of immobilities.[13]

Although this way of viewing motion is, I believe, logically impeccable, it may be psychologically difficult to accept. Perhaps the problem can best be seen in connection with the regressive form of the

Dichotomy paradox. Here we have Achilles at the starting point at the very moment at which the race begins. What, we ask, must he do first? Well, someone might say, first he has to run to the starting point of the tortoise. But that answer cannot be correct, for before he can do that, he must run to a point halfway between his and the tortoise's respective starting points. Before he can do that, however, he must get to a point halfway to the halfway point. And so on. We are off on the infinite regress. It seems that there is no first thing for him to do; whatever we suppose his first task to be, there another that must be completed before he can finish it. There is, in other words, no first interval for him to cross. This conclusion is true. But it does *not* follow that Achilles cannot get started.

Consider the arrow once more. Suppose it is at point C midway in its flight path. When we ask how it gets from C to B we may be wondering, consciously or unconsciously, where it goes next—how it gets to the next point. But this question is surely illegitimate, for we are thinking of the arrow's path as a continuous one. Since the points in a continuum are densely ordered, there is no next point. Between any two distinct points there is another (and, hence, infinitely many). The question about Achilles, which we just considered in connection with the regressive Dichotomy, may arise from the same psychological source. We may feel that his first act must be to get to the point next to his starting point, but no such point exists. According to the at-at theory of motion, this fact is no obstacle to motion. Both space and time are regarded as continuous, and hence, densely ordered. True, there is no next point of space for Achilles to occupy, but also there is no next moment of time in which he must do so. For each moment of time there is a corresponding point, and for each spatial point there is a corresponding moment; nothing more is required.

The psychological compulsion to demand a next point or a next moment may arise from the fact that we do not experience time as a continuum of instants without duration, but rather, as a discrete series of specious presents, each of which lasts perhaps a few milliseconds. Aside from anthropomorphism, however, there is no reason to try to impose the discrete structure of psychological time upon the mathematical notion of time as a continuum, since the continuous conception has proved itself such an extremely fruitful tool for the description of physical motion.[14]

Limits of functions. There is one final issue, arising out of the paradoxes of motion, that was significantly clarified by nineteenth-century foundations of mathematics. During the preceding two centuries, while the calculus floated on vague spatial and temporal intuitions, there was considerable controversy about the ability of a function to reach its limit. Some functions seemed to do so; others did not. It was all quite baffling. This puzzle relates directly to Zeno's paradoxes of Achilles and the Tortoise and the progressive form of the Dichotomy. Achilles seems capable of chasing the tortoise right up to the point of overtaking him, but can he reach that limiting point? Likewise, on the track by himself, Achilles seems capable of traversing the various fractional parts of the course right up to the finish line, but can he achieve that limit? Again, the definitions of functions and limits provided in the nineteenth century come nicely to the rescue. A limit is simply a number. A function is simply a pairing of two sets of numbers. If the limit happens to be one of the numbers in the set of values of the function, then the function does assume the limiting value for some value of its argument variable. If not, then the function never assumes the limiting value. No further question about the ability of a function to "reach" its limit can properly arise.

There can be no serious doubt that the aforementioned nineteenth-century mathematical developments went a long way in resolving the problems Zeno raised about space, time, and motion. The only question is whether there are any remaining problems associated with the paradoxes of motion. Beginning about 1950, a number of mathematically sophisticated writers, who were fully aware of the foregoing considerations, felt that an important problem still remained. One of the most articulate was Max Black, who argued that the analysis of Achilles' attempt to catch the tortoise into an infinite sequence of distinct runs introduces a severe logical difficulty.[15] The problem, specifically, is whether it even makes sense to suppose that anyone has completed an infinite sequence of runs. Black puts the matter forcefully and succinctly when he says that the mathematical operation of summing an infinite series will tell us where and when Achilles will catch the tortoise if he can catch the tortoise at all, but that is a big "if." There is, Black argues, a fundamental difficulty in supposing that he can catch the tortoise, for, he maintains, "the expression, 'infinite series of acts,' is self-contradictory."[16]

Black's argument is based upon consideration of a number of imaginary machines that transfer balls from one tray to another.[17] Suppose, for instance, that there are two machines, Hal and Pal, each equipped with a tray in front. When a ball is placed in Hal's tray, he moves it to Pal's tray; when a ball is placed in Pal's tray, he moves it to Hal's tray. They have a sort of friendly rivalry about getting rid of the balls. Suppose, further, that they are programmed in such a way that each successive transfer of the ball takes a shorter time; in particular, when the ball is first put into either tray, the machine takes 1/2 minute to move it to the other tray, next time it takes 1/4 minute, next time 1/8 minute, and so forth. (Actually, it is more like a frantic compulsion to get rid of the ball; they carry the maxim "It is more blessed to give than to receive" to a ridiculous extreme.) We begin by putting a ball in Hal's tray, and he takes 1/2 minute to move it to Pal's tray. Pal then takes 1/2 minute to put it back in Hal's tray, during which time Hal is resting. Then Hal takes 1/4 minute to transfer it to Pal's tray, while Pal is resting; in the next 1/4 minute Pal returns it to Hal's tray while Hal rests. As the process goes on, the pace increases until we see just a blur, but at the end of two minutes it is over, and both machines come to rest. The ball has been transfered infinitely many times; in fact, each machine has made infinitely many transfers (and enjoyed infinitely many rest periods) during the two minutes.

Now, we must ask, where is the ball? Is it in Hal's tray? No, it cannot be in Hal's tray, because every time it was put in Hal removed it. Is it in Pal's tray? No, because every time it was put there Pal removed it. Black concludes that the supposition that this infinite sequence of tasks has been completed leads to an absurdity.

Another hypothetical infinity machine—perhaps the simplest—is the Thomson lamp.[18] This lamp is of a common variety; it has a single push-button switch on its base. If the lamp is off and you push the switch, the lamp turns on; if the lamp is on and you push the switch, the lamp turns off. Now suppose that someone pushes the switch an infinite number of times; he accomplishes this by completing the first thrust in 1/2 minute, the second in 1/4 minute, the third in 1/8 minute, much as the runner in the Dichotomy is supposed to cover the infinite sequence of distances in decreasing times. Consider the final state of the lamp after the infinite sequence of switchings. Is the lamp on or off? It cannot be on, for each time it was on it was switched off. It cannot be off, for each time it was off it was switched on.

The speed of switching demanded is, of course, beyond human capability, but we are concerned with logical possibilities, not "medical" limitations. Moreover, there are mechanical difficulties inherent in the speed required of Hal and Pal as well as Thomson's lamp, but we are not concerned with problems of engineering. Further, there is no use trying to evade the question by saying that the bulb would burn out or the switch would wear out. Even if we could cover such eventualities by technological advances, there remains a logical problem in supposing that an infinite sequence of switching (or ball transfers) has been achieved. The lamp must be both on and off, and also, neither on nor off. This is a thoroughly unsatisfactory state of affairs.

Black and Thomson are *not* maintaining that Achilles cannot overtake the tortoise and finish the race. We all know that he can, and to argue otherwise would be silly. Black is arguing that it is incorrect to *describe* either feat as "completing an infinite sequence of tasks," and Thomson draws a similar moral. They are suggesting that the paradoxes arise because of a misdescription of the situation.

These authors have focused upon a fundamental point. We must begin by realizing that no definition, by itself, can provide the answer to a *physical* problem. Take the simplest possible case, the familiar definition of arithmetical addition of two terms. We find, *by experience*, that it applies in some situations and not in others. If we have m apples in one basket and n oranges in another, then we will have $m + n$ pieces of fruit if we put them together in the same container. (Popular folklore notwithstanding, we obviously can "add" apples and oranges.) However, as is well known, if we have m quarts of alcohol in one bucket, and n quarts of water in another, we will not have $m + n$ quarts of solution if we put them together in the same container. The situation is simply another instance of the relation between pure and applied mathematics discussed in the preceding chapter. We can define various mathematical operations within pure mathematics, but that is no guarantee of their applicability to the physical world. If such operations are to be applied in the description of physical facts we must determine empirically whether a given physical operation is an admissible interpretation of a given mathematical operation. We have just seen that the combining of apples and oranges in fruit baskets is a suitable counterpart of arithmetical addition, while the mixing of alcohol and water is not. A more significant example occurs in Einstein's special theory of relativity, where composition of velocities is seen not to be a physical counterpart of standard vector addition, as we shall see in the next chapter.

The same sort of question arises when we consider applying the (now standard) definition of the sum of an infinite series. Does a given physical situation correspond to a particular mathematical operation, in this case, the operation of summing an infinite series? Black concludes that the running of a race does not correspond to the summing of an infinite series, for the completion of an infinite sequence of tasks is a logical impossibility. Thus, the running of a race cannot correctly be described as completing an infinite sequence of tasks. This conclusion has far-reaching implications for modern science. If it is right, the usual scientific description of the racecourse as an infinitely divisible mathematical continuum is fundamentally incorrect. It may be a useful idealization for some purposes, but Zeno's paradoxes show that the description cannot be literally correct. The inescapable consequence of this view would seem to be that mathematical physics needs a radically different mathematical foundation if it is to deal adequately with physical reality.

Before accepting any such result, we must examine the infinity machines more closely. They do involve difficulties, but Black and Thomson have not identified them accurately. Consider Thomson's lamp. (The same considerations will apply to Black's infinity machines or any of the others.) Thomson has described a physical switching process that occupies one minute. Given that we begin at t_0 with the lamp off, and given that a switching occurs at $t_1 = 1/2$, $t_2 = 3/4$, and so on, we have a description that tells, for any moment *prior to* the time $T = 1$ (that is, one minute after t_0), whether the lamp is on or off. For $T = 1$, and subsequent times, it tells us nothing. For any time *prior to* T that the lamp is on, there is a subsequent time *prior to* T that the lamp is off, and conversely. But this does not imply that the lamp is both on and off at T; we can make any supposition we like without logical conflict. We have, in effect, a function defined over a half-open interval $0 \leqslant t < 1$, and we are asked to infer its value at $t = 1$. Obviously, there is no definite answer to such a question. If the function approached a limit at $t = 1$, it would be natural to extend the definition of the function by making that limit the value of the function at the end point. But the "switching function" describing Thomson's lamp has no such limit, so any extension we might choose would seem arbitrary.[19] The same goes for the position of the ball Hal and Pal pass back and forth. In the Dichotomy and the Achilles paradoxes, by contrast, the "motion function" of the runner does approach a limit, and this limit provides a suitably appealing answer to the question about the location of the runner at the conclusion of his sequence of runs.[20]

One cannot escape the feeling, however, that there are significant and as yet unmentioned differences between the infinite sequence of runs Achilles must make to catch the tortoise and the infinite sequence of ball transfers executed by Black's machines (or the infinite sequence of switch pushes required by the Thomson lamp). And there is at least one absolutely crucial difference. Consider the motion of the ball as it is passed back and forth between Hal and Pal. Say that the trays are three inches apart. Then the ball is made to traverse this *fixed* positive distance infinitely many times. In order to do so, it must travel an *infinite* distance in a finite length of time. Now, no one is interested in showing that Achilles can run an infinite distance in a finite amount of time—he is fast, but not that fast. The problem is to show how he can run a *finite* distance that can be subdivided into an infinite number of subintervals.

Achilles can make his run if he can achieve a fixed positive velocity; the ball which travels back and forth over the fixed distance between Hal and Pal must achieve velocities that increase without any bound. This difficulty could, of course, be repaired. Suppose we stipulate that the distances covered by the ball, like the distances Achilles must cover, decrease as the time available for each transit decreases. This can be done by making the trays of Hal and Pal move closer and closer during the two-minute interval, so that they coincide in the middle at the end of the infinite sequence of transfers. But now there is no problem at all about the position of the ball at the end—it is right in the middle in both trays! Similar considerations apply to the Thomson lamp. In order to accomplish a switching, the button must be moved a certain finite distance, say 1/8 inch. If this is done infinitely many times, the finger which pushes the button and the button itself must traverse an infinite total distance. A necessary, though not sufficient, condition for the convergence of an infinite series is that the terms converge to zero. In order to overcome this difficulty, the switch would have to be modified in some suitable way, in which case an answer can be given to the question regarding the final on-off state of the lamp.[21]

In the literature on Zeno's paradoxes of motion, especially that concerned with the infinity machines, a good deal of emphasis has been placed on the question of whether Achilles can be said to perform an infinite series of *distinct* tasks. When we divide up the racecourse into an infinite series of positive subintervals, it is often claimed, we are artificially breaking up what is properly considered one motion into infinitely many parts which—so the allegation goes—cannot be considered as individual tasks. In order to clarify this question, Adolf Grünbaum has given Achilles a fictitious twin—a doppelgänger—who runs a parallel racecourse, starting and finishing at the same time as the original Achilles.[22]

The new Achilles is a jerky runner. He starts out and runs the first half of the course twice as fast as his counterpart, and then stops and waits for him. When the slower one reaches the midpoint, the interloper runs twice as fast to the three-quarter mark, and again waits for the slower to catch up. He repeats the same performance for each of the remaining infinite series of subintervals. Grünbaum calls the original Achilles, who runs smoothly from start to finish, the *legato runner;* his new twin who starts and stops is called the *staccato runner*. The important facts about the staccato runner are: (1) He reaches the end of the course at the same time as the legato runner; if the original Achilles can run the course, so can the staccato runner. (2) The staccato runner takes a rest of finite (non-zero) duration between each of his infinite succession of runs; hence, there can be no question that he performs an infinite sequence of *distinct* runs. (3) The staccato runner (while he is running) runs at a fixed velocity which is simply twice that of his legato mate, so he is not involved in the kinds of ever-increasing velocities that were required in the unmodified Black and Thomson devices.

There is just one final feature of the staccato Achilles which might be a source of worry. Although he is not required to achieve indefinitely increasing velocities, he is required to do a lot of sudden stopping and starting, shifting instantaneously from velocity zero to velocity $2v$ (where v is the legato runner's velocity) and back again. This clearly involves infinite accelerations—and infinitely many of them. One could reasonably doubt the possibility of this degree of jerkiness. It turns out, however, that even the discontinuity in velocity is not a necessary feature of the staccato runner. The physicist Richard Friedberg has shown, by means of a complicated mathematical function, how to describe the motion of a more sophisticated (and less jerky!) staccato runner who covers *each* of the infinite sequence of subintervals by starting from rest, accelerating continuously to a maximum finite velocity, decelerating smoothly to rest, and remaining at rest for the required interval between runs. This staccato runner executes a motion conforming to a continuous function; his velocity (first derivative) and acceleration (second derivative) are continuous, as are all of the higher time-derivatives as well. Moreover, the peak velocities that occur in the successively shorter runs also decrease, converging to zero as the length of the run also converges to zero.[23] It is hard to see what kind of logical (or conceptual) objection can be raised against this kind of motion. But if the sophisticated staccato runner's series of tasks is feasible, so would be the motions of any of the appropriately modified infinity machines. The motion of the ball passed between Hal and Pal, for example, could be described by a combination of two such functions—the first would describe a sequence of motions from left to right with interspersed periods of rest; the second would consist of a similar sequence, but with the motions from right to left. The second set of motions would be executed during the periods of rest granted by the first function, and the first set of motions would occur during the rest periods granted by the second function. It therefore appears that a suitably designed Hal-Pal pair of infinity machines are logically possible if the *legato* Achilles—the one we all granted from the beginning—can complete his ordinary garden-variety run.

The most recent Zenoesque problem with which I am familiar is interesting partly because its Zenonian features are not immediately evident. It was originally published in *Mathematics Magazine* (January, 1971) as a straightforward mathematical puzzle, but it was picked up by Martin Gardner who recognized its affinities to Zeno's paradoxes of motion. He restated it in his "Mathematical Games" column in the *Scientific American* as follows:*

> A boy, a girl, and a dog are at the same spot on a straight road. The boy and the girl walk forward—the boy at four miles per hour, the girl at three miles per hour. As they proceed the dog trots back and forth between them at ten miles per hour. Assume that each reversal of directions is instantaneous. An hour later, where is the dog and which way is it facing?
>
> *Answer:* The dog can be at any point between the boy and the girl, facing either way. *Proof:* At the end of one hour, place the dog anywhere between the boy and the girl, facing either direction. Time-reverse all motions and the three will return at the same instant to the starting point.[24]

Here is the response I sent in reply to a letter from Gardner:

> Almost everyone has heard the old chestnut about the bird that flies back and forth between two approaching locomotives. Say that they are 30 miles apart, that each is traveling at 15 mph, and that the bird flies back and forth at 60 mph as they approach. How far does the bird fly before the two engines meet? Or, to achieve historical perspective, suppose that Achilles is pursuing the tortoise, and a Trojan fly buzzes back and forth between them. Given a set of velocities and distances, and our latter-day assurance that Achilles will overtake the tortoise at a determinate time and place, we can easily figure out how far the fly will travel. So far, we have no new Zenonian paradoxes. But, as Martin Gardner pointed out in a recent letter, a problem due to A. K. Austin of Sheffield University [stated above] brings up a new aspect of the old puzzle; in fact, it is just the time reversal of the bird and train problem.
>
> In order to retain historical perspective, let us go back to Achilles and the tortoise. Despite the initial handicap traditionally imposed upon Achilles, he catches the tortoise, and to redress the grievance he has long held against Zeno, he keeps on running, steadily increasing his lead over the fortunate tortoise. (I consider him fortunate in this version of the tale—at least in comparison with Lewis Carroll's account, "What the Tortoise Said to Achilles," in which Achilles stops and seats himself on the back of the tortoise, much to the latter's discomfort.) But consider the Trojan fly, who attempts to continue flying back and forth between the two runners even after the faster overtakes the slower. When Achilles and the tortoise are just even, the fly finds himself precisely in the position of Mr. Austin's dog.
>
> For the sake of definiteness, say that the tortoise travels at 1 mph, Achilles at 5 mph (he's been running since 500 B.C., so he is not as fleet as he once was), and the fly at 10 mph. They all arrive at the common meeting point without difficulty. But, to paraphrase Mr. Gardner's

---

*Gardner's statement and my response to it is from "Mathematical Games" by Martin Gardner, Dec. 1971. Copyright © 1971 by Scientific American, Inc. All rights reserved.

attested by the large number of letters received by *Mathematics Magazine* and by Martin Gardner from mathematically knowledgeable readers claiming that in some way or other the statement of Austin's boy-girl-dog puzzle contains an inconsistency.

A PARADOX OF PLURALITY

Although Zeno is best known for his four paradoxes of motion, he did propound a number of other paradoxes, including one that is even more fundamental. Although it is generally known as a *paradox of plurality*, it can plausibly be construed as a geometrical paradox which calls into question the very structure of the geometrical line (or any other continuum). Zeno presents the argument in terms of physical things and their parts, but the considerations he brings to bear seem to depend only upon the fact that these things are extended— that is, they occupy some finite, non-zero stretch of space. Although he talks about the possibility of subdividing the parts, he is not talking about the possibility of cutting up a physical object into separate physical parts that can be moved away from one another. He is not dealing with the physical hypothesis of the atomic constitution of matter. Rather, his arguments depend upon the possibility of making conceptual or mathematical divisions; for example, even if there are physical atoms (or subatomic particles) that cannot be split in two, if they occupy an extended region of space—be it ever so small—that space can be divided in the sense that we can distinguish its parts geometrically.

Since physical separation of parts is not at issue, we can just as well discuss the composition of the mathematical line. Zeno's argument runs as follows.[27] As we have seen from both the Achilles and Dichotomy paradoxes, any line segment is infinitely divisible. If we stop short with only a finite number of divisions, it is always possible to carry the division further. The process of halving the line, and then halving the half, is one which has no end. Hence, if the line is made up of parts, as it surely appears to be, then there are infinitely many of them. Now, Zeno poses a simple dilemma. What is the size of the parts? If they have zero magnitude, then no matter how many of them you add together, the result will still be zero. The process of adding zeroes never yields any answer but zero. If, however, the parts have a positive non-zero size, then the sum of the infinite collection of them will be infinite. In other words, a line segment must have a length of either zero or infinity; a line segment one inch or one mile long is impossible.

An immediate objection might be raised against the claim that the whole must have an infinite magnitude if the parts have non-zero size, since our discussion of the Achilles and the Dichotomy paradoxes showed how it is entirely possible for an infinite series of positive terms to have a finite sum. But this response is inappropriate here. In order for an infinite series of positive terms to converge, it is necessary that there be no smallest term; the sequence of terms must converge to zero. This condition clearly rules out the possibility of convergence for an infinite series of positive terms *all of which are equal to one another*. In the Achilles and the Dichotomy paradoxes we could rest content with the division of a line segment into unequal parts, for we were not trying to divide it up into its ultimate parts. It is hard to see, however, how different ultimate parts could have different sizes. If one "ultimate" part were larger than another, it would seem that the larger would be further subdividable, and hence not ultimate after all. Zeno apparently saw this point quite clearly.[28] So, the second horn of the dilemma still stands: if the (ultimate) parts have non-zero size, the whole is infinite in extent.

Let us look, then, at the first horn. We have already investigated the problem of adding up the infinitely many terms of an infinite series. We form the sequence of partial sums and, if it has a limit, we take that limit as the sum of the series. Obviously, an infinite series whose terms are all zero will converge to zero, since every partial sum, being a sum of a finite number of zeroes, will be equal to zero. It is small wonder that philosophers from Aristotle to Bergson have denied that the line is composed of points!

We have, however, left out a crucial fact. As Georg Cantor, the father of modern set theory, discovered toward the end of the nineteenth century, the number of points in a finite line segment is greater than the number of positive integers. Both numbers are, of course, infinite, but they are not equal to one another. The number of points in a finite line segment (or in the entire infinitely long straight line) is c (standing for *continuum*); the number of positive integers is $\aleph_0$ (pronounced "aleph null"; aleph is the first letter of the Hebrew alphabet). A set which has the same number of elements as the set of positive integers is said to be "denumerable" or "countably infinite." Sets with larger numbers of members, including, of course, those with c members, are called "non-denumerable" or "uncountably infinite." Infinite sequences and series involve countable sets of terms; they can be placed in one-to-one correspondence with the positive integers. It is impossible to establish a one-to-one correspondence between the set of positive integers and any set containing c elements. Hence, if we try adding up a non-denumerable number of zeroes we are stuck at the outset, since addition is defined only for sums of finitely many terms or sums of denumerable sets of terms. The operation of addition is not even defined for a non-denumerable set of terms; consequently, we have no justification for the conclusion of the first horn of the dilemma. We have no basis whatever for saying that the sum of a *non-denumerable* set of zeroes must be zero. This conclusion of Zeno does not follow.[29]

I do not mean to say that it is impossible to extend the definition of addition a step farther so as to make it applicable to sums of c terms, nor to claim that if this is done the sum of c zeroes must be other than zero. I imagine it is possible, consistently, to define such an operation in a way that yields the answer zero for the sum of any number of zeroes, finite, denumerable, or non-denumerable. Nevertheless, the mathematics we have considered so far does not force us to any such extension of the concept of addition, and for present purposes it would obviously be unwise to adopt one, if there is any way to avoid it.

There is a viable alternative, as Grünbaum has shown, in modern measure theory.[30] This theory provides a generalization of the concept of length of an interval. An interval is the set of points between two endpoints A and B. If the interval contains both end points, it is a closed interval $[A,B]$; if both end points are excluded, it is an open interval represented as (A,B); and if only one end point is included we have a half-open interval, either $[A,B)$ or $(A,B]$. For two fixed end points A and B, the measures assigned to all of these intervals are equal. This accords with our standard concept of length; the addition or removal of an end point does not change the length of the interval. Measures, like lengths, are additive. If an interval I is divided into two non-overlapping subintervals I_1 and I_2, the measure of I must be equal to the sum of the measures of I_1 and I_2. If A and B are not distinct points, but are one and the same, the interval $[A,B]$ (which is the same interval as $[A,A]$) is said to be degenerate—it is the unit set which contains only the point A. Because measures are additive, the degenerate interval must receive measure zero, and so must any set which contains any finite number of points. This measure is, moreover, extended to infinite sets of intervals, and it is said to be denumerably or countably additive. This means that the measure of an infinite sequence of non-overlapping intervals is equal to the sum of the measures of these intervals (where the sum is defined in just the way we have already explained for the summation of infinite convergent series). It is therefore possible to assign measures to denumerably infinite sets of points, and for all such sets the measure is zero. By a further extension, we can say that the entire Euclidean straight line is also an interval, whose measure is positive infinity ($+\infty$).

The most direct way—and also the standard way—of handling measures of point sets in a given line is to assign coordinates to the points of the line in the usual way. To each point is assigned a real number; the numerical difference between two coordinates is taken to be the distance between those two points, and also the length of the interval of which they are the end points. It is clear, for example, that the measure of all of the points between zero and one—the measure of the interval [0,1]—is equal to one. Since the set of all rational numbers is denumerable, the measure of all of the points in that interval with rational coordinates is zero. The measure of all of the irrational points between zero and one is, therefore, by subtraction, one. This set, of course, is not an interval, nor is it the union of any finite or denumerable set of intervals. It is doubtful that we could properly refer to the "length" of any such set; nevertheless, it does have a well-defined measure. We see that this concept of measure is, indeed, a generalization of the concept of length. However, not all sets of points on the line have measures; for reasons we need not go into, some sets are not measurable, and they receive no measure at all (which is *not* to say that their measure is zero, for zero is a very definite measure).

It is important to emphasize the fact that measure theory does not represent merely an extension of ordinary arithmetical addition (including the summation of infinite series) to the addition of non-denumerable sets of terms. In elementary arithmetic, if we are given a set of terms, say 2, 3, 5, it has the unique sum 10. Given the same set of terms again, the sum must be the same once more. In ordinary addition, even the order of the terms does not matter, but in dealing with infinite series the order of the terms may make a difference.[31] However, given the same infinite set of terms in the same order, the sum must always be the same. For example, our series

$$\frac{1}{2} + \frac{1}{4} + \frac{1}{8} + \ldots$$

has the unique sum 1, and the infinite series

$$0 + 0 + 0 + \ldots$$

has the unique sum 0.

Measures do not behave in the same way. As Cantor showed, any line segment of any length with its end points removed has precisely the same number of points as any other, and the infinite straight line also has the same number c. Moreover, the points composing any such open interval or entire line have precisely the same internal ordering amongst themselves. This can be shown by a simple diagrammatic argument (see Figure 5). Given two line segments AB and CD of unequal length, we may place the shorter above the longer and connect the end points of AB and CD with lines that intersect at point P. Using P as a point of projection, we can connect any point in

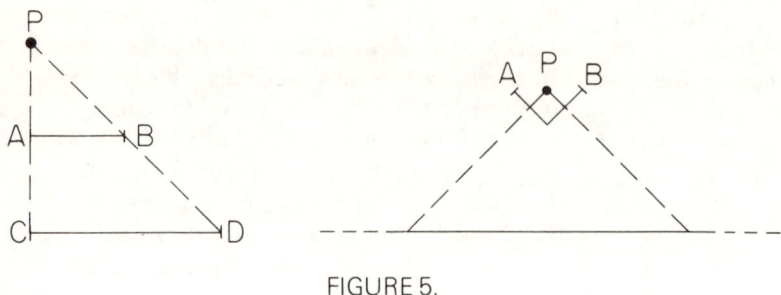

FIGURE 5.

the open interval (AB) to a point in the open interval (CD) by a line through P, and we can similarly connect any point in (CD) to a point in (AB). This shows that there must be the same number of points in (AB) and (CD), for we have just shown how to establish a one-to-one correspondence between the members of the two sets of points. By breaking the segment AB, we can show by similar reasoning that the open interval (AB) has the same number of points as the infinite line. Moreover, this correspondence between the points on the two lines is order-preserving; that is, if two points a and b in (AB) correspond respectively to two points c and d of (CD), then if a is to the left of b we will find c to the left of d. The existence of such an order-preserving one-to-one correspondence is the defining characteristic of sameness of order; two sets that have the same order in this precisely defined sense are said to be *isomorphic* to one another. Thus, we see that every open interval, finite or infinite, is isomorphic to every other.

It is an immediate consequence of these facts that the measure of an interval is not uniquely determined by the number of points it contains and the order in which they occur. Hence, if we assign each point measure zero, and attempt to "sum" them in the order in which they occur, we find that a given set of terms in a given order does *not* determine a unique "sum." The measure of a set of points depends upon more than the size (measure) of each of the points and the order in which they occur.

We have just seen that point sets containing c elements could have any finite length (measure) greater than zero, or infinite length. We have also seen that any set of points with a finite or denumerably infinite number of members must have zero length (measure). To prevent the tempting misconception that the measure of a set of points is greater than zero if and only if it has cardinality c, let us consider Cantor's ingenious *discontinuum*; it contains c points, but has measure zero. We begin with a line segment, say the set of points between zero and one, end points included. We remove the middle

A CONTEMPORARY LOOK AT ZENO'S PARADOXES

third of this line, but without taking the end points. This leaves the closed intervals [0, 1/3] and [2/3, 1]. Next we remove the open intervals that constitute the middle thirds of each of these intervals, leaving four closed intervals. This process is continued indefinitely, always removing the open middle third of each closed interval produced in the preceding stage. Pictorially, it looks like this:

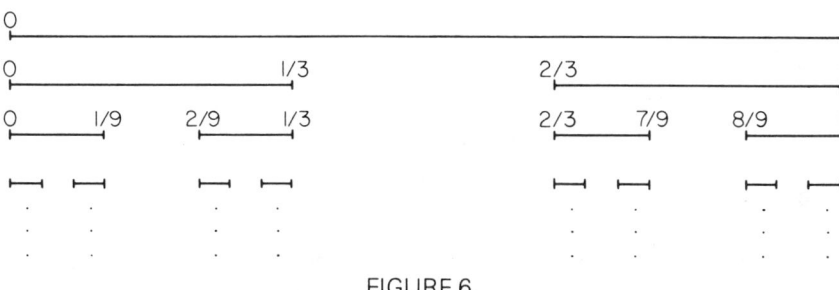

FIGURE 6.

The lengths of the open intervals removed at each stage form the following infinite series

$$\frac{1}{3} + \frac{2}{9} + \frac{4}{27} + \frac{8}{81} + \ldots + \frac{2^{n-1}}{3^n}$$

whose sum is 1. Having started with a segment with length 1, and having removed a denumerably infinite set of open intervals whose lengths add up to 1, we are left with a set of points having measure zero. This set is called a discontinuum because, between any two points remaining, at least one interval has been removed, and the remaining points still possess the cardinality of the continuum c.

To prove that the cardinality of the remaining set is c, we assign to each point on the segment [0, 1] its real number coordinate value, but we express this number in ternary notation—that is, in the notation that uses only the three digits "0," "1," and "2." In this notation, the points whose coordinates have "0" in the first place belong to the first third of the line, and those which have a "2" in the first place belong to the third third. The points with a "1" in the first place belong to the middle third, and they were removed. At the next stage, we removed the points from the middle of the first third and from the middle of the third third; these are the points whose coordinates have a "1" in the second place. The successive removals take away the points whose coordinates have "1" in the third, fourth, fifth, and so on, places. The result of the infinite sequence of removals is to get rid of all points that have the numeral "1" anywhere in the ternary expression of their coordinates. The remaining coordinates are expressed as infinite sequences of "0" and "2." And all possible sequences of this description remain.

In each of these sequences, we now change every "2" to "1." The result is a collection of sequences of "0" and "1," and these can be interpreted as binary representations of real numbers. Indeed, we have every possible sequence of "0" and "1," so there is one such sequence for each real number between zero and one. In this way we have established a one-to-one correspondence between the points in the Cantor discontinuum and the real numbers between zero and one. We know that this latter set has the cardinal number c. QED.

We have found that finite or denumerably infinite sets of points necessarily have measure zero in standard measure theory, but that sets of points of cardinality c (if they have any measure at all) may have measure zero, a positive finite measure, or a measure of positive infinity. This means that having cardinality c does not in any way determine the measure of a set of points. And we have seen, moreover, that for segments of non-zero measure (whether finite or infinite), there is no relation between measure and internal ordering of points, for all open intervals, regardless of length, share the same internal order. This means that *the measure of an interval or line segment is determined by the coordinate numbers we assign to the end points, and not by any fact about the internal structure of the interval or the way it is constructed out of its constituent points.* These considerations serve to resolve Zeno's "paradox of plurality," for they show how we can meaningfully assign non-zero finite measures to finite intervals and segments without running into any contradiction regarding the "addition of zeroes."

The foregoing findings enable us to shed more light on a major problem that was discussed in the preceding chapter—namely, the problem of ascertaining the geometrical structure of physical space. We maintained that the answer to this question depends in an essential way upon our interpretation of the concept of congruence of spatial intervals. In that context we argued that various definitions of congruence are admissible, some involving "universal forces," others not. At that point we called attention to a fundamental objection. In that context we claimed that it was equally legitimate to say that a particular measuring rod, which is not affected by any differential forces, shrinks to half its former size as it is moved from one place to another, or to say instead that it retains the same size wherever it is transported. Suppose, for instance, that our measuring rod is placed with its two ends coinciding with points A and B. It is then moved to a different position where its ends coincide with points C and D. We said that we are free to stipulate that interval AB is congruent to interval CD, or to stipulate that the length of AB equals say twice the length CD. But, you might object, whether we know how to ascertain the answer or not, the basic question is whether the interval AB contains *the same amount of space* as the interval CD. The answer is "yes" or "no." If it is "yes," then we are mistaken if we say that the rod changed its length upon being moved. A Newtonian, for instance, would insist upon a careful distinction between absolute space and various external devices (such as solid bodies) that are used to investigate its structure.

The foregoing analysis of measures of the continuum provides an answer to this Newtonian objection. We have seen that the standard measure of an interval depends upon the coordinate numbers assigned to its end points, not upon any intrinsic structure of the interval itself. Thus, interval AB received the same measure as interval CD whenever the corresponding coordinate differences were equal; otherwise, they received a different measure. But the assignment of coordinate numbers to points on a line is something we do by stipulation or convention. We cannot insist that coordinate numbers be assigned in such a way as to make equal coordinate differences coincide with equal distances, for equal distance is *defined* in terms of equal coordinate differences.

Grünbaum has summarized these considerations by describing the geometrical continuum as "metrically amorphous." Metrical amorphousness (as he characterizes it) depends upon two distinct factors: (1) the arrangement of the elements into the order type of the linear continuum, and (2) the qualitative homogeneity or indistinguishability of the elements. The colors, for example, can be arranged in a linear continuum, but the different colors—the distinct hues—are not qualitatively alike. The color continuum thus violates the second condition for metrical amorphousness. The points of space, by contrast, do not differ qualitatively. One might differ from another by being the site of a blue flash, while the other is the site of a pink flash, but these are qualitative differences between the events occurring at different places, not differences in the points in and of themselves. Grünbaum concludes, then, that space has no *intrinsic metric*; its internal structure does not determine distance relations.[32] We establish distance relations by the way we assign coordinates to points, or by the way we define the geometrical congruence relation.

We now have two distinct ways of establishing distance relations, and they might not always agree with one another. A word should be said to clarify the relations between them. For simplicity, let us deal solely with the relations among the points on a single line.

Given such a line (see Figure 7), let us assign coordinate numbers to its points, subject only to the condition that the coordinate numbers satisfy the same betweenness relations as the points to which they are assigned. In other words, if point B lies between points A and C on the line, then the coordinate assigned to B must be a number between the coordinate numbers of points A and C—for example, if B lies between A and C it would be impermissible to assign A the coordinate 1, B the coordinate 2, and C the coordinate 3/2. Over and above this condition, however, the assignment of coordinates is totally arbitrary. Both I and II (Figure 7) are acceptable ways of assigning coordinates to the points A, B, C, D, E.

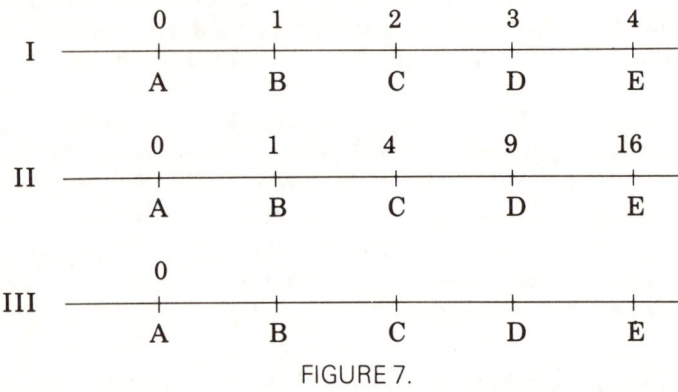

FIGURE 7.

The next step is to decide upon a method for determining the distance between two points when we know their coordinates; let us call such a method a *metric rule*. The most obvious metric rule is the one we employed above to assign a measure to the interval between two points: the distance between two points is the difference between their coordinates. Adopting this metric rule, we find that AB is congruent to CD in system I, since $1 - 0 = 3 - 2$. In system II, using the same metric rule, these intervals are not congruent, because $1 - 0 \neq 9 - 4$. While it is convenient to take the coordinate difference itself as a measure of the distance between two points, it is not necessary to do so. One could, for instance, define the distance as one-half of the coordinate difference, or as the difference between the squares of the coordinates of the end points. Using this latter rule in combination with coordinate system I is obviously tantamount to the use of coordinate system II in conjunction with the simpler standard metric rule. Clearly it is the combination of coordinate system and metric rule that determines the distance relations among the points on the line. The important fact is, given a system of coordinates and a metric rule, all congruence relations among intervals on the line have been determined.[33]

If a measuring rod is now introduced, we can ascertain empirically how it will behave. Suppose, for example, that the end points of the rod coincide with the points A and B, respectively, when it is situated on that part of the line, and with the points C and D, respectively, when it is located farther to the right. In the coordinate system I, adopting the metric rule according to which the length of the interval is equal to the difference of the coordinates of the end points, the rod remains the same size in those two different locations. Under these conditions, the rod is not subject to any universal forces that change its size when it is transported from one of these positions to the other. In coordinate system II, adopting the same metric rule, the rod expands from a length of 1 to a length of 5 as it is moved from the first position to the second. Under these conditions, the rod is subject to universal forces which result in a change in size. We could, of course, adopt a metric rule which equates the length of an interval to the differences between the square roots of the coordinates of the end points; with this rule and coordinate system II, we again find that the rod remains the same size when it is moved from one place to another. Which of these descriptions is correct? They are all correct, for they are equivalent descriptions of the same situation.

There is another way to attack the same line. Instead of assigning coordinates arbitrarily, we might use our measuring rod for this purpose. We could choose the point A as our origin, assigning it the coordinate 0, and mark off the intervals with our rod. Placing one end at A, we find as a matter of fact that the other end coincides with point B. Moving to the right, we find further that the right end of the rod coincides with point C when the left end is at B; moreover, the right end coincides with D when the left end is placed on C, and so on, as shown in III (Figure 7). Let us now *stipulate* that our measuring rod is free from universal forces—that it retains the same length wherever it is located. It follows that the intervals AB, BC, CD, DE, are all congruent to one another. If we assign the coordinates 1 to B, 2 to C, 3 to D, and 4 to E, adopting the metric rule which equates length with coordinate difference, we will express the mutual congruence of these intervals. We could, of course, have chosen a different coordinating definition, subjecting our rod to universal forces. We could have said that it expands to three times its original size when it is moved one position to the right, five times its original size when it is moved to the next position to the right, seven times its original size when it is moved to the next position, and so forth. If we assign the coordinate 1 to B, 4 to C, 9 to D, 16 to E, adopting the standard metric rule, we find that the intervals AB, BC, CD, DE are all incongruent to each other. Again, we have come up with equivalent descriptions of the same situation, both of which are equally correct.

We have found two ways of approaching the question of length of the intervals on a line. First, it was possible to assign coordinates arbitrarily and choose freely among metric rules. Then the question of whether our measuring rod stayed the same size, or changed in size, as it was moved around—the question of whether it was subject to universal forces—had to be settled by empirical investigation. Second, it was possible to stipulate, via a coordinating definition, whether the rod remained the same size or not as it was transported from place to place. When this was done, the question of which intervals on the line are congruent to one another had to be answered empirically. The situation reiterates the results of our discussion of Chapter 1. We said there that any geometrical description of the world has two components, a specification of congruence and a geometry. Either could be chosen freely, by stipulation; the other then became a factual matter (at least in part). Choosing the geometry, and then investigating for the presence or absence of universal forces, is parallel to stipulating the coordinate system and metric rule—the first of the foregoing two approaches. Choosing a coordinating definition of congruence, and then ascertaining the geometrical structure of space, is parallel to the use of the measuring rod to provide the combination of coordinate system and metric rule. In no case, however, does the internal structure of the line as an ordered set of points dictate its metrical structure. Its metrical character is imposed from without by such means as choice of coordinate system, choice of metric rule, or behavior of some sort of measuring instrument. Our investigation of the structure of the linear continuum thus reinforces our earlier results regarding our freedom to select alternative definitions of congruence.

THE DISCRETE VS. THE CONTINUOUS

The infinitesimal calculus has long been—and still is—the basic mathematical tool in the description of physical reality. It employs variables that range over continuous sets of values, and the functions it deals with are continuous. Although the calculus has been completely "arithmetized," so that its *formal* development does not demand any geometrical concepts, it is still applied to phenomena that occur in physical space. Its applicability to spatial occurrences is achieved through analytic geometry, which begins with a one-to-one correspondence between the points on a line and the set of real numbers. The set of real numbers constitutes a continuum in the strict mathematical sense; consequently, the order-preserving one-to-one correspondence between the real numbers and the points of the geometrical line renders the line a continuum as well. If, moreover, the geometrical line is a correct representation of lines in physical space, then physical space is likewise continuous. Motion is treated, moreover, as a function of a continuous time variable, and the function itself is continuous. The continuity of the motion function is essential, for velocity is regarded as the first derivative of such a function, and acceleration as the second derivative. Functions which are not continuous are not differentiable, and hence they do not even have derivatives. Continuity is buried deep in standard mathematical physics. It is for this reason that we have concerned ourselves at length with the problems continuity gives rise to.[34]

A serious objection might be raised, however, to the view that the mathematical continuum provides a precise and literal representation of physical reality. Since physics customarily uses such idealizations as frictionless planes, point-masses, and ideal gases, the argument could go, it might be reasonable to suppose that the mathematical continuum is another idealization that is convenient for some purposes, but does not provide a *completely* accurate description of space, time, and motion. There is, in addition, ample precedent for treating magnitudes that are known to be discrete as if they were continuous. The law of radioactive decay, for example, employs a continuous exponential function even though it is universally acknowledged that the phenomenon it describes involves discrete disintegrations of individual atoms. Where very large finite numbers of entites are involved, the fiction of an infinite collection is often a convenient one which yields good approximations to what actually happens. In electromagnetic theory, for another example, the infinitesimal calculus is used extensively in dealing with charges, even though all the evidence points to the quantization of charges. It has sometimes been suggested that these considerations hold the solution to Zeno's paradoxes. For instance, the physicist P. W. Bridgman has said, "With regard to the paradoxes of Zeno . . . if I literally thought of a line as consisting of an assemblage of points of zero length and of an interval of time as the sum of moments without duration, paradox would then present itself."[35]

Although I am in complete agreement with the claim that physics uses idealizations to excellent advantage, it does not seem to me that this provides any basis for an answer to Zeno's paradoxes of plurality or motion. The first three paradoxes of motion purport to show a priori that motion, if it occurs, must be discontinuous. Indeed, Zeno's intention, as far as we can tell, seems to have been to prove a priori that motion cannot occur. With the exception of a very few metaphysicians of the stripe of F. H. Bradley, most philosophers would admit that the question of whether anything moves must be answered on the basis of empirical evidence, and that the available evidence seems overwhelmingly to support the affirmative answer. Given that motion is a fact of the physical world, it seems to me a further empirical question whether it is continuous or not. It may be a very difficult and highly theoretical question, but I do not think it can be answered a priori. Other philosophers have disagreed. Alfred North Whitehead believed that Zeno's paradoxes support the view that motion is atomistic in character, while Henri Bergson seemed to hold an a priori commitment to the continuity of motion.[36] It seems to me that considerable importance attaches to the analysis of Zeno's paradoxes for just this reason. Space and time may, as some physicists have suggested, be quantized, just as some other parameters, such as charge, are taken to be.[37] If this is so, it must be a conclusion of sophisticated physical investigation of the spatio-temporal structure of the atomic and subatomic domains. A priori arguments, such as Zeno's paradoxes, cannot sustain any such conclusion. The fine structure of space-time is a matter for theoretical physics, not for a priori metaphysics, physicists and philosophers alike notwithstanding. The result of our attempts to resolve Zeno's paradoxes of motion is not a proof that space, time, and motion are continuous; the conclusion is rather that for all we can tell a priori it is an open question whether they are continuous or not.

Before we finally leave Zeno's paradoxes, something should be said about the view of space, time, and motion as discrete quantities. The historical evidence suggests that some of Zeno's arguments were directed against this alternative; that is a plausible interpretation of the Stadium paradox at any rate. Zeno seems to have realized that, if space and time both have discrete structure, there is a standard type of motion that must always occur at a fixed velocity. If, for instance, an arrow is to fly from position A to position B in as nearly continuous a fashion as is possible in discrete space and time, then it must occupy adjacent space atoms at adjoining atoms of time. In other words, the standard velocity would be one atom of space per atom of time. To travel at a lesser speed, the arrow would have to occupy at least some of the space atoms for more than one time atom; to travel at a greater speed, the arrow would have to skip some of the intervening space atoms entirely, never occupying them in the course of the trip. All of this sounds a bit strange, perhaps, but surely not logically contradictory; this is the way the world might be. Moreover, it is possible, as Zeno's original Stadium paradox shows, for two arrows to pass one another traveling in opposite directions without ever being located next to one another. Imagine two paths, located as close together as possible in our discrete space, between A and B. Let one arrow travel one of these paths from A to B, while the other travels the other path from B to A (see Figure 8). Suppose that the arrow traveling the upper track leaves A and occupies the first square on the left, while the arrow traveling the lower track leaves B at the same

A CONTEMPORARY LOOK AT ZENO'S PARADOXES

(atomic) moment of time, occupying the first square on the right end of his path. Let each arrow move along its track at the rate of one square for each atom of time. At the fourth moment, the upper arrow is just to the left of the lower arrow; at the next moment, the upper arrow is just to the right of the lower arrow. At no moment are they side-by-side—they get past one another, but there is no event which qualifies as the passing (if we mean being located side-by-side traveling in the opposite directions). This is strange perhaps, but again, it is hardly logically impossible.

A	1	2	3	4	5	6	7	8	B
	8	7	6	5	4	3	2	1	

FIGURE 8.

The mathematician Hermann Weyl has, however, posed a basic difficulty for those who would like to quantize space.[38] If we think of a two-dimensional space as being made up of a large numbers of tiles (something like Figure 8), we get into immediate trouble over certain geometrical relations. Suppose for example, that we have a right triangle ABC in such a space (see Figure 9). Consider, first, the

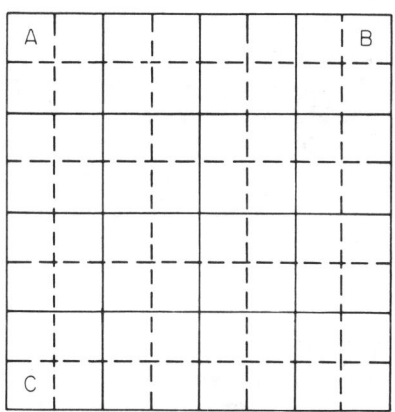

FIGURE 9.

tiles drawn with solid lines. If the positions A, B, and C represent the respective corner tiles, then we see that the side AB is four units long, the side AC is four units long, and the hypotenuse BC is also four units long. The Pythagorean theorem says, however, that the square of the hypotenuse equals the sum of the squares of the other two sides. This means that a right triangle with two legs of four units each should have a hypotenuse about 5-2/3 units long. The Pythagorean theorem is at least approximately true in physical space, as we have found by much experience. The result based upon tile-counting does not begin to approximate the correct result.

This example shows something important about approximations. It is easy to see that discontinuous motion in discrete space and time would be difficult to distinguish from continuous motion if our space and time atoms were small enough. It might be tempting to suppose that our geometrical relations would approach the accustomed ones if we make our tiles small enough. This, unfortunately, is not the case, as you can see by taking the finer grid in Figure 9 given by the broken and solid lines together. Instead of 16 tiles, we now have 64 tiles covering the same region of space. But looking at our triangle ABC once more, we see that all three sides are now 8 units long. No matter how small we make the squares, the hypotenuse remains equal in length to the other two sides. No wonder this is sometimes called the "Weyl tile" argument![39] This is one case in which transition to very small atoms does not help at all to produce the needed approximation to the obvious features of macroscopic space. It shows the danger of assuming that such approximation will automatically occur as we make the divisions smaller and smaller.

It is important to resist any temptation to account for the difficulty by saying that the diagonal distance across a tile is longer than the breadth or height of a tile, and that we must take that difference into account in ascertaining the length of the hypotenuse of the triangle. Such considerations are certainly appropriate if we are thinking of the tiles as subdivisions of a continuous background space possessing the familiar Euclidean characteristics. But the basic idea behind the tiles in the first place was to do away with continuous space and replace it by discrete space. In discrete space, a space atom constitutes one unit, and that is all there is to it. It cannot be regarded as properly having a shape, for we cannot ascribe sizes to parts of it—it has no parts.

Now, I do not mean to argue that there is no consistent way of describing an atomic space or time. It would be as illegitimate to try to prove the continuity of space and time a priori as it would be to try to prove their discreteness a priori. But, in order to make good on the claim that space and time are genuinely quantized, it would be necessary to provide an adequate geometry based on these concepts. I am not suggesting that this is impossible, but it is no routine mathematical exercise, and I do not know that it has actually been done.[40]

SUGGESTED READINGS

1. Benardete, José. *Infinity; An Essay in Metaphysics.* Oxford: Clarendon Press, 1964.
2. Boyer, Carl B. *The History of the Calculus and its Conceptual Development.* New York: Dover Publications, Inc., 1959. Previously published under the title, *The Concepts of the Calculus.*
3. Courant, Richard, and Robbins, Herbert. *What is Mathematics?* New York: Oxford University Press, 1941.
4. Grünbaum, Adolf. *Modern Science and Zeno's Paradoxes.* Middletown, Conn.: Wesleyan University Press, 1967. British edition, somewhat revised. London: George Allen & Unwin Ltd., 1968.
5. Salmon, Wesley C., ed. *Zeno's Paradoxes.* Indianapolis: The Bobbs-Merrill Co., 1970. An anthology containing a number of important modern discussions and an extensive bibliography.

Chapter Three

A TRIP ON EINSTEIN'S TRAIN

There is no need to recount the history of astronomy from antiquity to modern times, especially the overthrow of Ptolemy's geocentric system by the heliocentric system of Copernicus, Galileo, and Kepler. It is an oft-told tale, and it has been told well by others, especially in the year 1973, which was the 500th anniversary of the birth of Copernicus—an occasion celebrated throughout the world. Nor is there any need to detail the development of Newtonian mechanics and its triumphant successes in explaining wide varieties of phenomena. When, in the latter part of the nineteenth century, Maxwell enunciated a unified theory of electromagnetic phenomena (including the propagation of light), which was soon confirmed experimentally by Hertz, it seemed as if classical physics had well-nigh achieved fundamental completeness and perfection. Few physicists realized that there was a profound difficulty in reconciling Newton's mechanics with Maxwell's electrodynamics, and that a revolution which would shake physics to its very foundations was just over the horizon at the turn of the century. Einstein, of course, played a vital role in that revolution.[1]

By the close of the nineteenth century, two characteristics of light seemed firmly established. First, light is a wave phenomenon. This feature had been dramatically confirmed early in the century when the great French mathematician Poisson—himself an opponent of the wave theory of light—deduced as a consequence of that theory that a bright spot should appear in the center of a shadow cast by a circular disk. He regarded this demonstration as a *reductio ad absurdum* of the wave theory of light, but lo and behold, when the experiment was performed, the bright spot which had never before been noticed was there (Figure 1). It is now a familiar phenomenon, known as the "Poisson bright spot," in honor of the man who, ironically, was confident that it did not exist. There was no way in which the contending corpuscular theory of light could explain this strange fact.

FIGURE 1. The Poisson Bright Spot. This figure shows the diffraction pattern of a small disk with the bright spot in the center, as predicted by Poisson on the basis of the wave theory of light. Such patterns arise from the interference of waves; they are inexplicable on the hypothesis that light is entirely corpuscular in nature.

Second, light waves travel with a constant velocity c, which is approximately 186,000 miles or 300,000 kilometers per second. This fact follows from James Clerk Maxwell's electromagnetic theory. According to his theory, an accelerating electric charge will set up a wave that propagates with this velocity. Further investigation revealed that there is a broad spectrum of electromagnetic waves all the way from radio waves many meters in length down to gamma rays whose lengths are less than a billionth of a millimeter. This spectrum includes, going from longer to shorter wave lengths (or smaller to greater frequencies), standard AM radio broadcast waves, FM radio and VHF television waves, UHF television waves, radar and microwaves, infrared (heat) radiation, visible light from red to violet, ultraviolet rays, X-rays, and gamma rays. All travel with the characteristic velocity c. The term for this velocity appears in Maxwell's equations, which govern all electromagnetic phenomena.

If these waves are all propagated with the velocity c, the natural question is, *velocity c relative to what?* The natural answer is, *relative to the medium through which the waves are propagated.* Consider sound waves for purposes of comparison. Sound consists of waves in air, and their velocity is measured relative to the air which transmits them. If, for example, atmospheric conditions are such that sound travels at 700 miles per hour, then that is the speed of propagation you will measure *if you are at rest relative to the air*. If you are traveling on a train that is moving at 100 miles per hour through the air in the same direction as the sound wave, you will find the velocity of the sound wave—relative to you—to be 600 miles per hour. If you are in an airplane traveling at 500 miles per hour in the same direction, you will find that the sound wave travels at 200 miles per hour—relative to you. If you are on a supersonic plane traveling at 1700 miles per hour, then the velocity of the sound wave—relative to you—is 1000 miles per hour in the opposite direction (that is, -1000 miles per hour).

It is a well-known fact that light is not transmitted by air as its medium, for light travels to us from the sun and distant stars through space that contains no air. Moreover, if a bell is mounted (by a suspension that does not conduct sound) in a glass jar from which all of the air is then removed, the ringing of the bell cannot be heard, but the bell remains fully visible. Thus, light travels through the vacuum, even though sound does not. (Indeed, when we speak of the velocity of light we shall always mean its velocity in a vacuum.) But since a wave motion seems to demand something that does the vibrating, nineteenth-century physicists postulated the "luminiferous ether" as a medium through which the light rays are propagated. This ether was thought to pervade all space. It cannot be pumped out of a container in the way that air or any other ordinary substance can. The velocity c which appears in Maxwell's equations is thus presumed to stand for the velocity of light, as well as all other electromagnetic radiation, *relative to the ether*.

If space is filled with this ether then—given that the earth moves through space in its annual revolution around the sun—the earth must be traveling through it. The American experimentalist A. A. Michelson devised an ingenious method for measuring the velocity of the earth relative to the ether. In 1887, he and his colleague, E. W. Morley, first performed the now-famous Michelson-Morley experiment. Although this experiment was precise enough to detect a motion only a hundredth of the supposed velocity of the earth through the ether, no motion was detected. This experiment was repeated at various seasons of the year, so that if at some particular time the earth happened to be virtually at rest in the ether, there should be detectable motion six months later when the orbital motion of the earth has changed by approximately 60 kilometers per second. Nevertheless, at no time was any motion relative to the ether detected. Michelson was totally mystified by the "failure" of this experiment.

We need not describe the details of the Michelson-Morley experiment, nor the efforts of classical physicists to explain the null result.[2] For our purposes it is sufficient to note that the point of departure for Einstein's special theory of relativity, first published in 1905, is the impossibility in principle of detecting our motion relative to the luminiferous ether.[3] If such motion had been detectable, then a frame of reference fixed at rest with respect to the ether would have been a very special frame of reference—namely, the frame of reference with respect to which the velocity of light through a vacuum has the special value c. Given our inability to detect any such privileged frame of reference, Einstein made the incredibly bold assumption that the whole idea of a unique ether frame of reference is dispensable, as is the concept of the ether itself. Instead, he postulated, the speed of light has the same value c when measured in *any* inertial frame of reference.[4] This amounts to saying that all of the laws of nature, including Maxwell's equations which contain the velocity constant c, hold equally in all inertial frames of reference. (The special theory of relativity is distinguished from Einstein's general theory of relativity (1916) in that the special theory pertains only to physical phenomena in inertial reference frames. A brief remark about the general theory will be found in the Epilogue of this book.)

Let us see just what this means. Assume for the present that the earth is an inertial reference frame. Suppose that a beacon fixed to the earth sends a light beam along a straight railroad track, and that someone farther down the track measures the speed with which the light beam is traveling past him. He finds that the speed has the value c which appears in Maxwell's equations. Very well. Now, suppose someone who is traveling on a train that moves at a very high velocity down the track also decides to measure the speed of that same light beam relative to his moving reference system (attached to the train). According to Einstein's postulate, he too must find that the speed of light has the very same velocity as was measured by the observer on the ground—namely c. Thus, whether the train is traveling at one-tenth the speed of light, or one-half the speed of light, or three-fourths the speed of light, the observer on the train will always get the same result when he measures the speed of the light beams sent out by the beacon situated on the ground. To appreciate how contrary this assumption is to common sense, recall the differing speeds measured for a sound wave, depending upon the state of motion of the observer relative to the medium (air) through which the sound wave is propagated. Nevertheless, Einstein introduced the assumption—which lies at the very foundation of his special theory of relativity—that the speed of light is the same in every inertial frame. Indeed, in his autobiography, he recalls thinking about this idea as a boy of sixteen. What would a light wave look like, he asked himself, if I pursued it at very high speed? The answer, it seemed to him, was just what Maxwell's theory said: it would look like a wave traveling along at velocity c. And at the time he also realized an immediate consequence—it is impossible in principle to chase a light ray at the speed of light.[5]

It was not until several years later—when he was twenty-five—that Einstein worked out a systematic theory on the basis of this assumption. He then realized that the new theory would demand a radical revision of certain basic concepts of physics and common sense, especially our concepts of space and time. The most fundamental revision would involve the concept of *simultaneity*—that is, of events happening at the same time. When events happen at the same place, simultaneity presents no particular problems. If the professor sneezes just as the class bell rings, we can observe that these two events happen simultaneously. This is a case of *local* simultaneity. But when the events in question happen at widely separated places, for example, on earth and on Jupiter, the problem of whether they occur at the same time is much more difficult.

In order to see what is involved in the problem of *distant* simultaneity, let us consider a famous "thought experiment" proposed by Einstein.[6] Imagine a train traveling at 6/10ths c (speed of light) along the track described in Figure 2. One observer (the "ground observer") is stationed on the embankment beside the track; another observer (the "train observer") is stationed at the middle of the train. Suppose that two lightning bolts strike the train, one at each end, each leaving a burn mark on the train and on the ground where it strikes (line *a*). Suppose, further, that the ground observer sees these two lightning bolts simultaneously—that is, the light from the two bolts as they strike the ground reach the eyes of the ground observer at the same time (line *c*). He measures the distance between himself and each of the marks on the ground left by the two lightning bolts; he finds that he was located precisely at the midpoint between them. He therefore says that the two lightning bolts struck at these two spatially separated points *at the same time*, for light rays traveling the same distance at the same velocity reached him at the same time. This, according to Einstein, is a suitable definition of simultaneity for two events which occur at different places: light rays emanating from those two events will reach the midpoint between them at the same time.[7]

Now, consider what the man on the train observes. He, too, will see the two lightning bolts strike, for light rays from these events will reach his eyes. But since he is traveling toward the light ray emanating from the bolt that strikes the front of the train, and away from the one that strikes the rear of the train *as viewed from the*

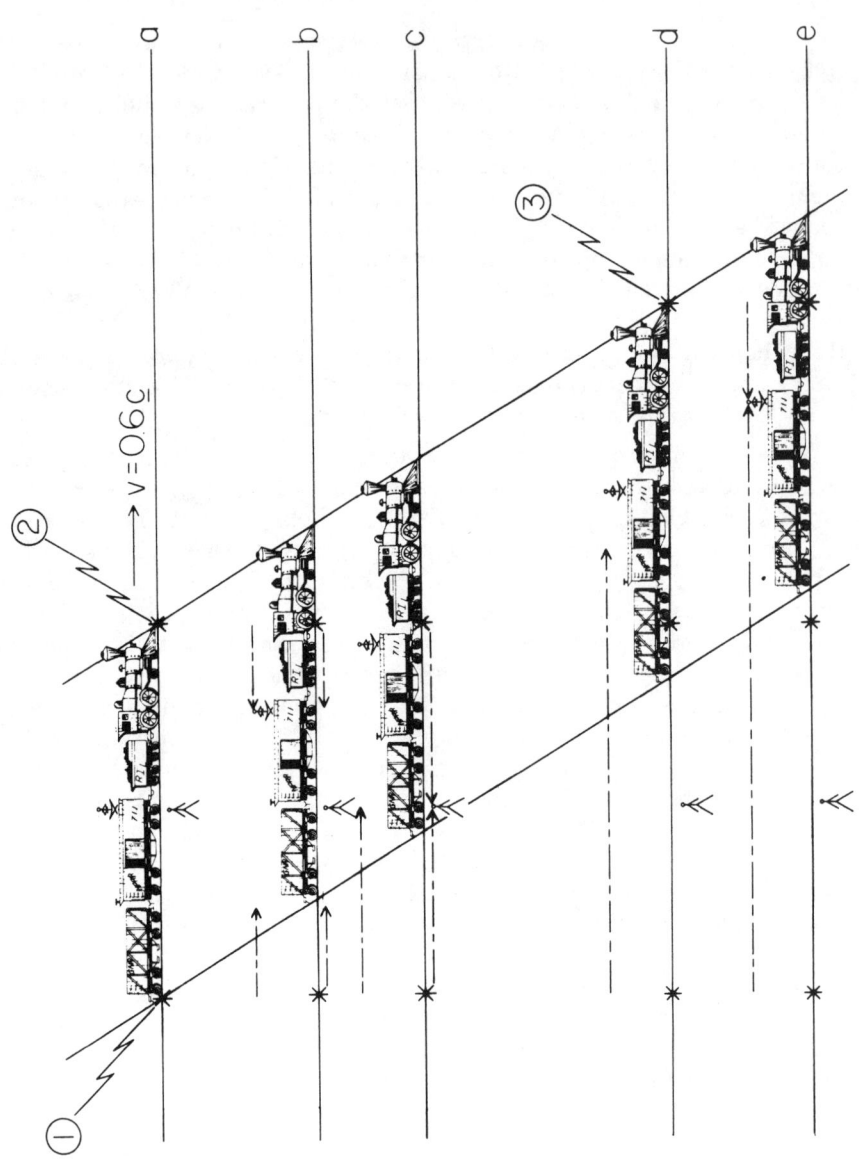

A TRIP ON EINSTEIN'S TRAIN

FIGURE 2. The train is traveling toward the right at six-tenths of the speed of light with reference to the observer standing on the ground beside the tracks.

a. Two bolts of lightning, 1 and 2, strike the rear and the front of the train, respectively, leaving burned marks (represented by stars) on the ground and on the train. One observer stands on the ground halfway between the two places where the lightning strikes the ground; another observer sits on the train halfway between the two places where the lightning strikes the train.

b. As the train moves rapidly to the right, the train observer sees the light from bolt 2 before the light from bolt 1 reaches him. He concludes that bolt 2 struck the train before bolt 1. Light from neither bolt has reached the ground observer yet.

c. Light from both bolts reaches the ground observer at the same time. He concludes that bolts 1 and 2 struck the ground simultaneously.

d. A third lightning bolt (3) strikes the front of the train, before the light from bolt 1 has reached the observer on the train. Bolt 3 strikes the train at exactly the same place as 2.

e. The light from bolt 1 and the light from bolt 3 both reach the observer on the train at the same time. He concludes that bolts 1 and 3 struck the train simultaneously, since they struck the train at equal distances from him, and the light from them travels toward him at the same speed.

The "snapshots" are unequally spaced because they are not separated by equal time intervals (cf Figure 4). The reader is reminded that Einstein first published his special theory of relativity in 1905.

ground system, he will see the lightning bolt strike the front of the train (line *b*) before he sees the other lightning bolt strike the rear of the train. Since he is also situated at the midpoint between the two events (he is in the middle of the train), he must declare that the two lightning bolts did *not* strike simultaneously, for the light rays emanating from the two events did *not* reach him at the same time. In his frame of reference, the locomotive was hit by lightning before the caboose. This, then, is the fundamental result: events occuring at different places which are simultaneous in one frame of reference will not be simultaneous in another frame of reference which is moving with respect to the first. This is known as *the relativity of simultaneity*. It will play a major role in the next chapter.

Although we have described these phenomena in terms of observers situated in these two frames of reference, the ground and the train, it would be a serious misunderstanding to suppose that there is anything subjective or illusory about the results. Each observer reports accurately on the events that occur; the results would be the same if inanimate instruments of observation replaced the human observers. Moreover, it is essential to remember that the striking of these two lightning bolts are the same two events regardless of which reference system is used to describe them. Simultaneity is relative, but that is an *objective* fact.[8] Relative to the ground system, bolts 1 and 2 really are simultaneous; relative to the train system, bolt 2 really does strike before bolt 1.

Suppose, further, that a third lightning bolt (see line *d* in Figure 2) strikes the front of the locomotive somewhat later than bolt 2 struck. Bolt 3 is simultaneous with bolt 1 in the train system (line *e*), although they are not simultaneous in the ground system. These are the objective facts of the situation.

At this point, for the purpose of clarifying both the present discussion and the following chapter, it will be helpful to introduce standard space-time diagrams—often called "Minkowski diagrams." These are two-dimensional diagrams in which one axis represents time and the other represents one of the spatial dimensions. Such diagrams are well-suited to represent situations like Einstein's train, for the motion of the train is spatially one-dimensional (since it moves along a straight line). Figure 3 is such a diagram.

In Figure 3, the horizontal axis, labeled x, represents the one spatial dimension of our frame of reference; in Einstein's train example, it represents the track along which the train moves from left to right. The vertical axis, labeled t, represents the history of one point which remains fixed (the origin) in that frame of reference. This axis

A TRIP ON EINSTEIN'S TRAIN

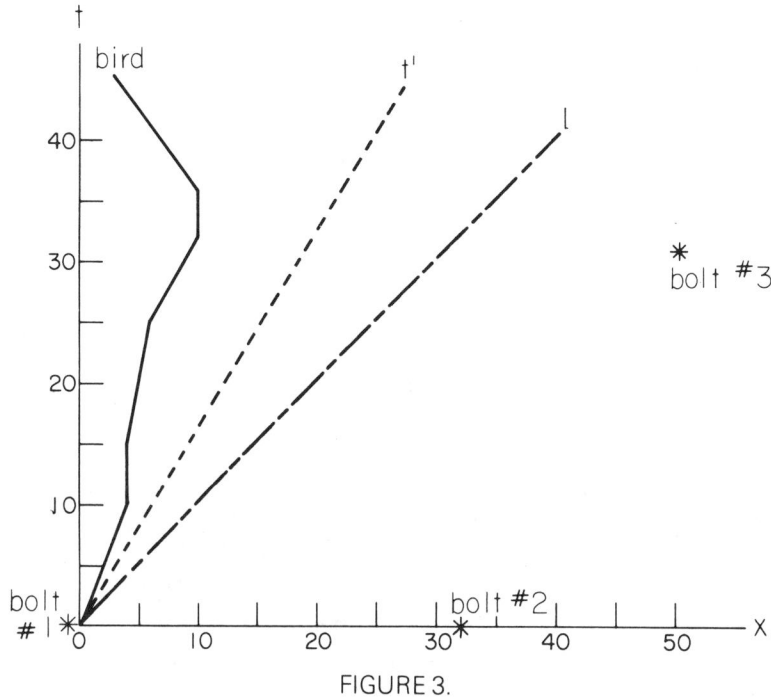

FIGURE 3.

could be taken to stand for the location at which lightning bolt 1 struck the ground in the train example. The different points on the t-axis represent different times at the same place. The different points on the x-axis represent different places at the same time. An event is an occurrence which happens at some particular place at some particular time; events are represented by points on the space-time diagram. The stars on the x-axis, for example, represent the events consisting of the lightning bolts striking the rear and the front of the train and the ground (bolts 1 and 2 in line a, Figure 2).

A physical thing is an object which exists throughout a span of time; the history of any particular physical object can be represented by a line in the space-time diagram. Lines which represent physical objects are called "world lines" of these objects. Suppose, for instance, that there is a rock on the ground where lightning bolt 1 struck. We may let the t-axis in Figure 3 be the world line of that rock—an object which remains at rest in the ground frame of reference.[9] The crooked line might be the world line of a bird which leaves the rock as the lightning strikes, moving to the right during the next ten units of time until it arrives at a point four units of distance away from the rock. There it perches during the next five units of time. It subsequently moves off farther to the right, first at a lesser speed and then at a greater speed, until it comes to rest again at a point ten units to the right of the rock. After resting there for a while, it begins its return flight back to the rock.[10] Obviously, the slope of the world line indicates the velocity with which the bird moves relative to the ground observer's frame of reference; when the world line is parallel to the t-axis the bird is at rest in the ground system. Larger deviation from the parallel orientation represents higher velocity.

We have been assuming that the reference frame of the ground observer constitutes an inertial system. Any system that is moving at uniform velocity with respect to an inertial system is itself an inertial system. Thus, the system of the moving train is an inertial reference frame. An object moving with uniform velocity in our ground frame of reference is represented by a straight line in the space-time diagram. The dashed line labeled t' is just such a line; in fact, it is the world line of the rear end of the train (the part struck by lightning bolt 1). It has been labeled t' in analogy to the t-axis on the ground observer; the t'-axis is the world line of an object which is at rest in the reference frame of the train.

We have yet to say anything about the units in which distances and time intervals are to be measured. For distance we can choose any of the customary units; let's say that it is to be measured in meters. For time measurement we shall adopt a unit that is rather unusual, but which is very convenient for discussing special relativity. This unit, which may be called a "light-meter" of time, is simply the amount of time it takes light to travel one meter.[11] Compared with everyday time units, such as the hour, minute, or second, the light-meter is a very small unit. Because light travels at approximately 300,000 kilometers per second (300,000,000 meters per second), the light-meter is 1/300,000,000th of a second. The result of this choice of unit for the space-time diagram is that the path of a light ray in space-time is represented by a straight line inclined at a 45° angle (the broken dash-dot line labeled l for "light" in Figure 3). This is an obvious feature of our choice of units; they were purposely chosen so that light will travel across one unit of space (one meter) during one unit of time (one light-meter).

Let us now use this equipment to diagram the "thought experiment" of Einstein's train; the result is Figure 4. To begin, we select the ground system as our rest system, and consequently make the world line of the ground observer a vertical line. The horizontal x-axis is the spatial axis. It represents simultaneity in the ground system, for it represents all of the different positions along the railroad track at the same time. Moreover, any other line parallel to the x-axis also exhibits a relation of simultaneity *in the ground frame of*

A TRIP ON EINSTEIN'S TRAIN 79

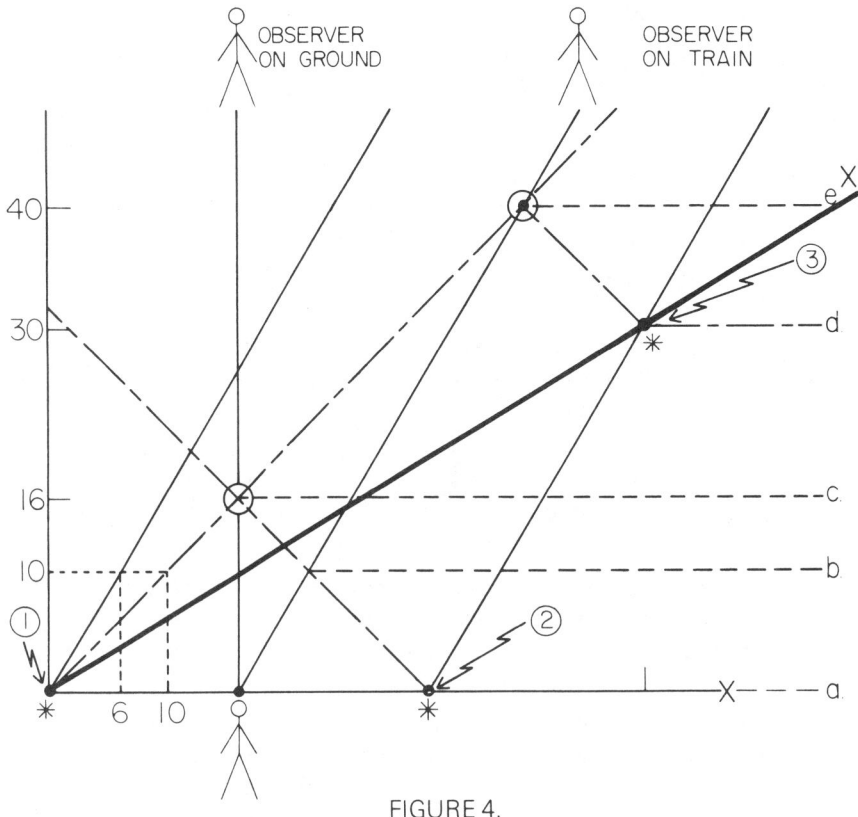

FIGURE 4.

reference. Since, as we have seen, simultaneity in the train system is not the same as simultaneity in the ground system, we shall have to investigate carefully the proper way to represent simultaneity in the train system. This is, in fact, the whole point of the exercise.

As we saw in analyzing Figure 2, the two lightning bolts 1 and 2 struck the two ends of the train simultaneously with respect to the ground system; the two stars on the x-axis equidistant from the ground observer represent these two lightning bolts. The letters a, b, c, d, e at the right of Figure 4 correspond to parts a, b, c, d, e of Figure 2. It will be advisable to compare Figures 2 and 4 throughout the discussion of Figure 4.

The dot-dash lines emanating from the points at which the two lightning bolts struck on the x-axis (drawn at 45°) represent the light rays which travel from the lightning bolts toward the two observers. In Figure 4, as in Figure 3, the world line of the rear of the train is an oblique straight line. In Figure 4 we have added two other oblique lines, both parallel to the world line of the rear of the train. The center line is the world line of the train observer; the other is the world line of the front of the train. As we see, lightning bolts 1 and 2 struck the two ends of the train just as the world lines of the two ends of the train coincide with the points at which the two bolts struck in the ground system. As we stipulated at the beginning, the train is traveling at 6/10 the speed of light. This fact is reflected in the angle at which the world lines of the parts of the train system are drawn. The train travels a distance of six units in the time it takes a light ray to travel a distance of ten units.

Next, we note (at b) that the light ray from bolt 2 reaches the train observer. This happens well before the light ray from bolt 1 reaches him.

A little later (at c) light rays from both bolts reach the ground observer at the same time; this event is represented by the circle on his world line. The simultaneity, *in his frame of reference*, of the striking of bolts 1 and 2 is thereby established.

At d (30 units later than a, relative to the ground system) lightning bolt 3 strikes the front of the train. This happens when the train is a distance of 18 units (relative to the ground system) to the right of its position when it was struck by bolt 2.

At e (40 time units later than a, relative to the ground system) the light from lightning bolt 3 reaches the observer on the train. At precisely the same moment, the light from lightning bolt 1 reaches him. This event is represented by a circle on the world line of the observer on the train. He therefore declares the striking of bolt 3 simultaneous with the striking of bolt 1. Therefore, the spatial axis, the x'-axis, which represents relations of simultaneity in the reference frame of the moving train, must pass through the two points defined by the striking of lightning bolts 1 and 3. This line is, of course, at a slant with respect to the horizontal x-axis of the ground system.

In Figure 5 we construct a diagram similar to Figure 3 by removing many of the complexities of Figure 4, in order to show in clear relief our essential results. The three stars, labeled 1, 2, and 3, represent the three events consisting of the striking of the three respective lightning bolts. The t-axis again represents the rock on the ground which was struck by lightning bolt 1; it is, of course, at rest in the ground frame. The x-axis again represents the locus in space-time of all events simultaneous, *with respect to the ground system*, with the striking of lightning bolt 1. It contains the event consisting of the striking of bolt 2.

The t'-axis again represents the world line of the rear end of the train; it is, of course, a point which is at rest in the reference frame of the train. Any world line parallel to the t'-axis would represent

A TRIP ON EINSTEIN'S TRAIN

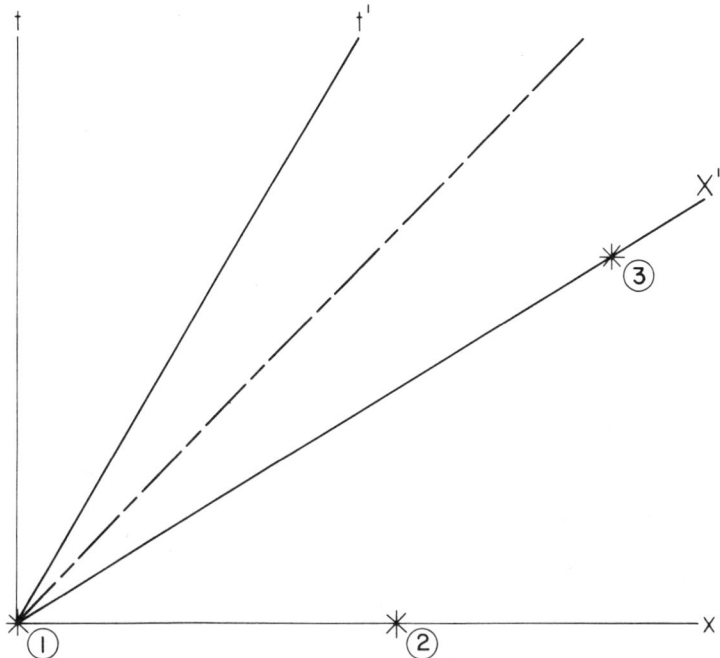

FIGURE 5. Simultaneity in moving reference frame.

a point at rest in the train system; for example, the world line of the train observer is parallel to the t'-axis. The x'-axis represents the space-time locus of all events simultaneous, *with respect to the train system*, with the striking of lightning bolt 1. The event consisting of the train being struck by lightning bolt 3 lies on the x'-axis. Any straight line parallel to the x'-axis defines a set of events all of which are simultaneous with each other with respect to the reference frame of the train.

Throughout classical physics, it was assumed that relations of simultaneity were absolute and unambiguous. Two events that happened at different places were either simultaneous or not—one's state of motion had nothing to do with the question. Einstein showed that simultaneity is relative to the particular reference frame. The spatial and temporal coordinates of a moving reference frame—not just the spatial coordinates—differ from those of a stationary reference frame. When we redefine rest, by changing to a moving reference frame, we must also redefine simultaneity.

Let us now consider another basic revision of our time concepts that is demanded by the special theory of relativity. This innovation is the so-called "time dilation" phenomenon; it gives rise to the "twin paradox" which is the point of departure for the next chapter. Time dilation can be seen immediately by considering the light clock, whose operation depends upon the special properties of light.

When we introduced the light-meter as the unit of time, we said nothing about how this unit is to be measured. Let us now consider that question. It is easy to devise a clock—known as a light clock—that measures time in light-meters. The light clock consists of two parallel mirrors mounted exactly one meter apart. A small light pulse (a photon[12]) is reflected back and forth between the two mirrors. One unit of time is the time it takes for the light pulse to travel from one mirror to the other. Since it is a basic postulate of the special theory of relativity that the speed of light is a constant, this clock can always be counted upon to give accurate time intervals, since the time taken for a trip of constant distance at constant velocity must be a constant time interval. The crucial fact will be that such a clock can be used to measure time in any inertial reference frame, because *the speed of light is the same in all inertial frames.*

Suppose that the observer aboard Einstein's train is equipped with a light clock. To the observer on the train the light clock is just as described: it consists of two mirrors parallel to each other, exactly one meter apart, with a light pulse bouncing back and forth between them. It is at rest in the train system. One unit of time (a light-meter) is the time it takes for the pulse to travel from one mirror to the other. Now, consider what the ground observer sees as he watches the light clock moving by him on the train (Figure 6). We assume that the clock is held in a vertical position, so that its orientation is perpendicular to the direction in which the train is traveling. It is evident that the path of the light pulse is not straight up and down with respect to the ground frame, but back and forth at a slant, as shown in Figure 6.

One fact about the moving light clock is immediately evident: as the light pulse travels from mirror to mirror, it travels a longer distance d than does the pulse in the clock that is stationary in the ground system (which travels the distance b). Since, according to our basic postulate, the speed of the light pulse in the moving clock is equal to the speed of the light pulse in the stationary clock, the unit of time in the moving system must be expanded or dilated with respect to the unit of time in the stationary system. Indeed, it is a matter of very elementary geometry to calculate the amount of this dilation. All we need is the Pythagorean theorem—the square of the hypotenuse of a right triangle is equal to the sum of the squares of the other two sides:

$$d^2 = a^2 + b^2 \qquad (1)$$

A TRIP ON EINSTEIN'S TRAIN

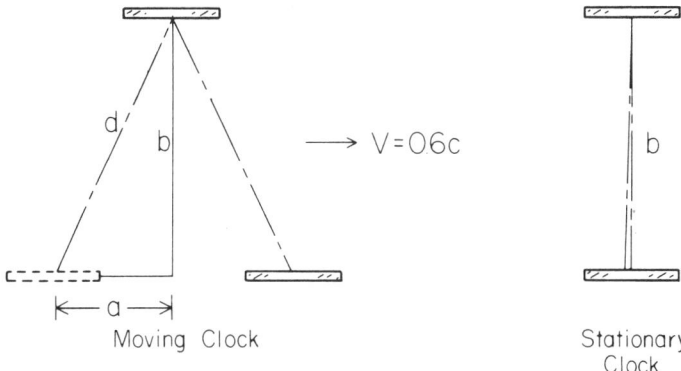

FIGURE 6. Light clock.

Let us denote the time interval required for the light pulse to travel from the lower to the upper mirror by Δt. Since the light pulse travels that distance at the velocity c,

$$d = c\Delta t \quad \text{and} \quad d^2 = c^2(\Delta t)^2 \qquad (2)$$

During the same time interval Δt, the light clock itself travels the distance a at velocity v; hence,

$$a = v\Delta t \quad \text{and} \quad a^2 = v^2(\Delta t)^2 \qquad (3)$$

Substituting the values from (2) and (3) into (1), we have

$$c^2(\Delta t)^2 = v^2(\Delta t)^2 + b^2 \qquad (4)$$

or rearranging,

$$c^2(\Delta t)^2 - v^2(\Delta t)^2 = b^2 \qquad (5)$$

Factoring the left side yields

$$(\Delta t)^2 (c^2 - v^2) = b^2 \qquad (6)$$

Dividing by $c^2 - v^2$ (which obviously is not zero), we get

$$(\Delta t)^2 = \frac{b^2}{c^2 - v^2} \qquad (7)$$

Dividing numerator and denominator of the right side by c^2 yields

$$(\Delta t)^2 = \frac{b^2/c^2}{c^2/c^2 - v^2/c^2} \qquad (8)$$

Taking the square roots of both sides, we have

$$\Delta t = \frac{b/c}{\sqrt{1 - v^2/c^2}} \tag{9}$$

Since our light clock defines the time unit, the amount of time required for the light pulse to travel a distance b at velocity c is one; hence,

$$\Delta t = \frac{1}{\sqrt{1 - v^2/c^2}} \tag{10}$$

This quantity is the time dilation factor. Since the denomenator is a positive quantity less than 1, the whole quantity is larger than 1. As we said at the outset, Figure 6 makes it obvious that the moving clock ticks off its units more slowly than the stationary clock. Its units, measured by the stationary clock, are longer than those of the stationary clock. This effect constitutes a retardation of the moving clock relative to the stationary clock. We shall discuss this retardation phenomenon at length in the next chapter.

Since we have specified the velocity at which Einstein's train is moving, let us compute the amount of retardation that will be experienced by the clocks on the train. Since $v = 0.6c$, we see that

$$\Delta t = \frac{1}{\sqrt{1 - .36c^2/c^2}} = \frac{1}{\sqrt{.64}} = \frac{1}{0.8} = 1.25 \tag{11}$$

Hence, each light-meter on the train equals one and one-fourth light-meters on the ground; each second on the train equals 1.25 seconds on the ground; each hour on the train equals 1.25 hours on the ground; each year on the train equals a year and a quarter on the ground. Since all clocks must be affected in the same way—otherwise there would be a way of picking out a privileged reference frame by comparison of other kinds of clocks with light clocks—all processes aboard the train will be slowed by the same factor. Hence a person riding the train for eight years of train time will age only eight years during the time interval that occupies ten years of ground time. This is the basis of the "twin paradox" that constitutes the point of departure for the next chapter.

According to Einstein's principle of relativity, there is no way of selecting a privileged reference frame which is *the* uniquely stationary frame (at rest, say, with respect to the luminiferous ether). We may, consequently, regard the train system as the stationary reference frame, and consider the ground system as the moving system. Looking at the matter this way, the light clock of the ground observer is the moving clock, and it will look to the train observer just like the moving clock in Figure 6, with the one difference that it is moving in the opposite direction. All of the computations we made when we figured the dilation factor for the light clock on the train as the moving clock will go through in precisely the same way if we substitute $-v$ for v throughout. But since the square of the velocity appears in the dilation formula, the sign does not make any difference. Thus, according to the special theory of relativity, the time dilation is completely reciprocal. Given two inertial reference frames in motion with respect to one another, *the time of each is dilated with respect to that of the other*. In the next chapter we shall consider the "clock paradox"—a challenge to the very logical consistency of this reciprocal time dilation phenomenon. When we carry through the detailed analysis, we shall see that the relativity of simultaneity plays an indispensable role in avoiding a logical contradiction.

Let us, finally, examine the way in which the length of an object depends upon its state of motion. The result will emerge directly from the time dilation phenomenon. Suppose that a telegraph line runs along beside the railroad track, and that it is supported by poles located 60 meters apart, as measured by the ground observer, using his meter stick. The ground observer knows that the train is traveling at $6/10\ c$; he knows that it will take the train observer 100 light-meters *of time* to travel from one pole to the next, since the train is traveling at 0.6 meters of distance per light-meter of time. To the observer on the train, the telegraph poles are rushing past at a velocity of $0.6c$ in the opposite direction.[13] He can use his clock to ascertain how long it takes for him to travel from one pole to the next. Applying what we have just established regarding time dilation, we see that he will measure the time to be $8/10$ of 100 light-meters, or 80 light-meters of time. Knowing the velocity and the time, he can easily compute the distance:

80 light-meters × 0.6 meters/light-meter = 48 meters

The distance between telegraph poles, which is 60 meters with respect to the ground system, is 48 meters in the reference frame of the moving train. This shortening of the length of a moving object or spatial distance is usually known as the *Lorentz contraction*.[14]

The method by which we have ascertained the length contraction in this special case can be generalized. Suppose we have any two inertial frames of reference in motion with respect to each other. Let us choose one of these frames and call it the "stationary" frame.

This choice is arbitrary; it does not matter which frame we designate in this way. The other frame is then called the "moving frame." Let v be the velocity with which the moving frame travels relative to the stationary frame. Let d be the distance, as measured in the stationary frame, between two points (A and B) in the stationary frame. Let Δt be the time interval, as measured in the stationary frame, for an observer at rest in the moving frame to travel the distance d between the two points. These quantities are related to one another in the familiar way:

$$v = d/\Delta t \tag{12}$$

We now introduce a standard condition—the *reciprocity condition*—which stipulates that the velocity at which the stationary frame travels relative to the moving frame is $-v$ (that is, equal in amount but opposite in direction to the velocity of the moving frame relative to the stationary frame). The status of this condition will be discussed in the next chapter. Let $\Delta t'$ be the time interval, as measured in the moving frame, required for the observer at rest in the moving frame to pass the same two points A and B in the stationary frame. Note that the passage of the observer in the moving frame past the two distinct points A and B are two objective events which can be observed and described by observers in both systems. We are concerned to compare the spatial distances and temporal intervals involved, as described in each of the two systems.

Our investigation of the time dilation phenomenon revealed that the time of the moving system is expanded by a factor of $1/\sqrt{1 - v^2/c^2}$, and consequently, the relation between the two time intervals Δt and $\Delta t'$ is given by

$$\Delta t' = \Delta t \sqrt{1 - v^2/c^2} \tag{13}$$

Since the sign of the velocity makes no difference to the amount of time required to travel from one of these points to the other we can write

$$v = d'/\Delta t' \tag{14}$$

where d' is the distance between A and B relative to the moving system. Consequently, (12) and (14) yield

$$d/\Delta t = d'/\Delta t' \tag{15}$$

It follows immediately that

$$d/d' = \Delta t/\Delta t' \qquad (16)$$

and hence,

$$d' = d\sqrt{1 - v^2/c^2} \qquad (17)$$

This is the standard equation for the Lorentz length contraction. It follows from the time dilation equation (which we extracted directly from consideration of the light clock) and the reciprocity condition.

As we said, the choice of which frame is to be considered stationary is completely arbitrary; we could just as well have selected the one we call "moving" to be the stationary system. All derivations would have worked out in precisely the same manner—except for the sign of the velocity which is, as already remarked, immaterial to the problem. We must conclude, therefore, that lengths in the stationary system are contracted relative to the moving system in precisely the same way as lengths in the moving system are contracted relative to the stationary system. This symmetry of length contraction is obviously parallel to the symmetry of time dilation; indeed, it is obvious from the way in which we derived the length contraction relation from the time dilation relation that the one symmetry would entail the other. We conclude, therefore, that given two inertial frames in motion with respect to one another, lengths parallel to the direction of motion in each are contracted when measured by observers at rest in the other.

The reciprocity of length contraction gives rise to a result which at first sight seems rather strange, if not downright paradoxical. Suppose that Einstein's train, as measured by an observer traveling on the train, is 100 meters. Suppose further, that the train passes through a tunnel which is 100 meters long, as measured by the ground observer. We ask each observer whether the train is ever completely inside of the tunnel. The ground observer says, "Yes, for the length of the tunnel is 100 meters, whereas the contracted length of the train as measured in my system, is 80 meters; hence, in order to pass through the tunnel the train will, for a time, be totally within it." The observer on the train has a different story. "No," says he, "for the length of the train is 100 meters, while the contracted length of the tunnel, as measured in my frame of reference, is 80 meters; hence, the tunnel is not large enough to contain the entire train." We must staunchly resist any temptation to attribute these apparently conflicting results to illusions on the part of the observers.

The example of the train and the tunnel shows the intimate relationship between the length contraction and the relativity of simultaneity. To ask whether the train is entirely within the tunnel amounts to asking a question about simultaneity—namely, are the two ends of the train ever within the tunnel *at the same time?* Since, as we already know, the relation of simultaneity in the train system differs from the relation of simultaneity in the ground system, it is not surprising that the two observers give two different answers. Suppose that two assistant ground observers are stationed within the tunnel, each one 10 meters from an end (see Figure 7). The chief ground observer stands at the middle of the tunnel. Each assistant ground observer has instructions to set off a small explosive charge at the precise moment when the respective end of the train reaches his location: the observer near the right end sets off his charge when the front of the train arrives at his point of observation; the observer on the left sets off his charge when the rear of the train arrives at his point of observation. The chief observer, stationed in the middle, watches to see whether the light from the flashes of the two explosions reach him at the same time. He observes that they do. He therefore concludes that the two ends of the train were, in fact, within the tunnel simultaneously.

Indeed, this operation *defines* the length of the moving object. To determine the length of an object that is in motion relative to a frame of reference, we determine the simultaneous positions of the two ends at some moment, and then measure the distance between them in our stationary frame of reference. In this case, obviously, the length of the moving train, measured within the ground observer's stationary reference frame, is 80 meters.

Suppose, further, that there are two poles, located outside of the tunnel at either end, each exactly 12½ meters from its respective end of the tunnel as measured in the ground system (Figure 7). In that system the two poles are, therefore, 125 meters apart. The chief observer in the train system holds his position at the midpoint of the train, but there are, in addition, two assistant observers on the train, one at each end. They, too, are equipped with small explosive charges. They are instructed to set off these charges as they reach their respective poles: the observer at the head of the train is to set off his charge when he reaches the pole outside of the far end of the tunnel; the observer at the rear end of the train sets off his charge when he reaches the pole outside of the near end of the tunnel. The chief observer notes that the light flashes from each of these charges reach him simultaneously, so he concludes that the two charges were exploded simultaneously. He therefore reports that the front end of the train emerged from the tunnel before the rear end entered the tunnel—the train was never wholly within the tunnel. If asked how far apart are the poles, as measured in his reference system, he answers 100 meters, for that is the distance in his system between the two ob-

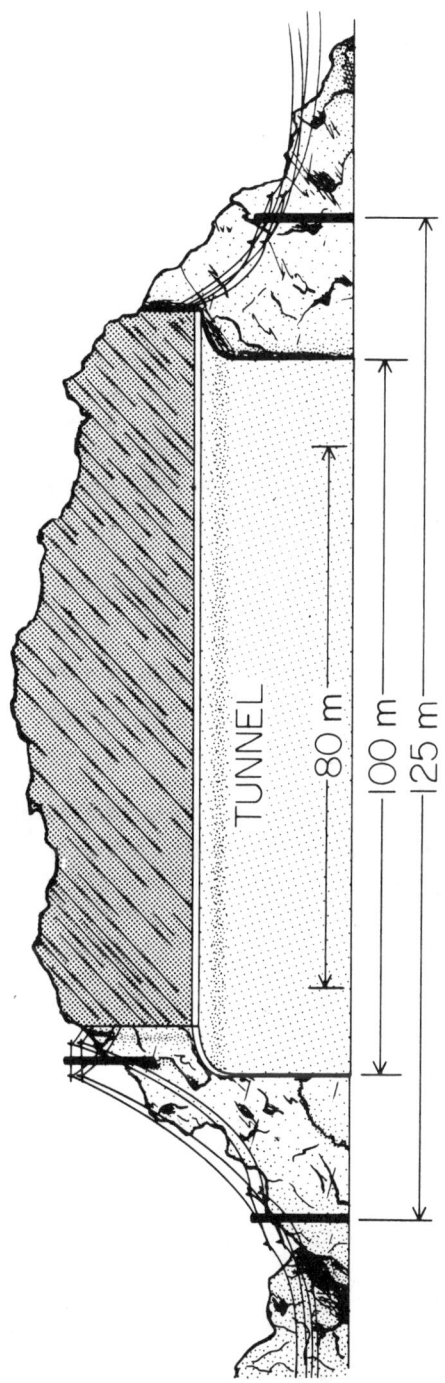

Figure 7. Reciprocity of length contraction.

servers at the opposite ends of the train. The observer in the train system ascertains the distance between the two poles which are at rest in the ground system by taking a simultaneity projection of that stretch of ground, and measuring the length of the projection (which happens to coincide with the length of the train) in his frame of reference.

As we noted at the outset, special relativity prohibits any physical object from being accelerated to the speed of light. Thus, if a rocket were launched with a velocity of 0.6c from Einstein's train, which is traveling with velocity 0.6c relative to the ground system, the rocket would not be traveling at 1.2c relative to the ground. In order to ascertain the velocity of the rocket relative to the ground, Einstein's "composition of velocities formula" must be used. If v is the velocity of the train relative to the ground, and u is the velocity of the rocket relative to the train, then V, the velocity of the rocket relative to the ground, is given by

$$V = \frac{u + v}{1 + \frac{uv}{c^2}} \qquad (18)$$

As long as u and v are both less than c, it always turns out that $V < c$. This formula will be used in the next chapter.

There is no need, in the present context, to discuss at length the empirical and experimental confirmations of the special theory of relativity. Suffice it to say that this theory is by now a standard item in the physicist's tool kit, and it is used daily in physical laboratories all over the world. A striking instance lies in the fact that the Stanford Linear Accelerator, which accelerates electrons to very nearly the speed of light, is approximately two miles long, and was constructed at enormous expense. If, however, classical Newton-Maxwell physics were correct and no relativistic effects were present, the same electron velocities would be achieved in less than one inch![15] Some additional empirical evidence will be mentioned in the next chapter.

SUGGESTED READINGS

1. Born, Max. *Einstein's Theory of Relativity.* New York: Dover Publications, Inc., 1962.
2. Einstein, Albert. *Relativity, The Special and the General Theory.* New York: Henry Holt and Co., 1921. Paperback ed., Crown Publishers, 1968.
3. Einstein, Albert, et al. *The Principle of Relativity.* New York: Dover Publications, Inc., n.d.
4. Taylor, Edwin F., and Wheeler, John Archibald. *Spacetime Physics.* San Francisco: W. H. Freeman and Co., 1966.

Chapter Four

CLOCKS AND SIMULTANEITY IN SPECIAL RELATIVITY or WHICH TWIN HAS THE TIMEX?*

One of the most surprising and highly publicized consequences of Einstein's theory of relativity is the so-called "twin paradox." If one of a pair of twins were to take a very fast trip in a rocket ship, while his brother remains on earth, the traveling twin upon returning would be much younger than his stay-at-home brother. This phenomenon has never been directly observed, however, even in the space travels of astronauts, because their velocities are much too small compared with that of light to produce a noticeable difference in human aging. Nevertheless, by using the extreme accuracy of modern atomic clocks, the difference in "aging" has been confirmed even at the speeds which are usual for conventional jet airliners nowadays.

In October, 1971, J. C. Hafele[1] placed several cesium beam clocks aboard ordinary commercial around-the-world jet flights. One flight circumnavigated the earth traveling in an eastward direction; the other in the opposite direction. These clocks were compared with similar clocks that remained at the U.S. Naval Observatory during the trip (see Figure 1). Relative to the Naval Observatory clocks, the clocks that traveled eastward lost 59 ± 10 nanoseconds (1 nsec = 10^{-9} sec = 1 billionth sec), while those traveling westward gained 273 ± 7 nanoseconds. These results are in excellent agreement with the theoretical predictions of standard relativity theory; according to Hafele and Keating, they "provide an unambiguous empirical resolution of the famous clock 'paradox'[2] with macroscopic clocks."[3]

This outcome is not in the least disturbing; it is if anything reassuring, especially as it bears upon the special theory of relativity. The time dilation effect had long since been dramatically confirmed by

*This chapter is reprinted from *Motion and Time, Space and Matter: Interrelations in the History of Philosophy and Science*, edited by Peter Machamer and Robert G. Turnbull. Copyright © 1975 by the Ohio State University Press.

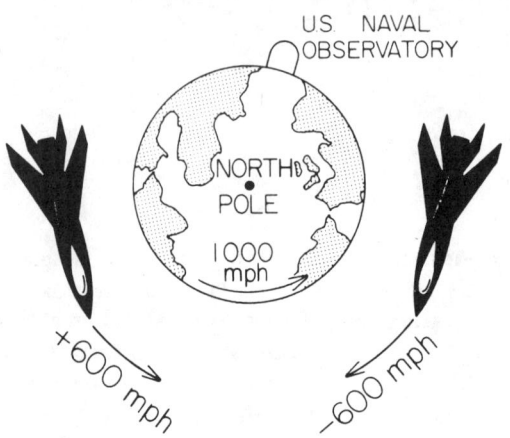

FIGURE 1. Twin paradox.

muon decay phenomena.[4] It is nice to see that macroscopic clocks exhibit the same retardation effects as do microscopic clocks; it would have been a great surprise if they had not.[5] Thus, we can say with even greater confidence than before that the twin who takes a very fast trip on a rocket ship will return younger than the twin who remains at home. There is nothing contradictory or unintelligible about this fact; at worst it is in the category of "strange but true."

Hafele's traveling clocks were, of course, transported through gravitational fields. Their behavior, consequently, depends in part upon general relativistic considerations, though special relativistic effects were also involved. In this discussion it is the special theory which is my object of concern, and it is the special theory which has been charged with outright inconsistency in connection with the traveling twin. For, it has been argued, if the first twin travels from A to B and back again, he will be younger than the second twin who remained at A. But, the argument continues, since there is no such thing as absolute rest, we could just as well say that the first twin remained at rest while the second twin (whom we previously characterized as remaining at A) has traveled away from the first twin and then returned to rejoin him. On this way of looking at the situation, however, the second twin should be younger than the first (see Figure 2)! There is nothing paradoxical in saying that one twin is younger than the other (twin paradox), but it is blatantly inconsistent to say that each is younger than the other when they meet after the trip (clock paradox).[6]

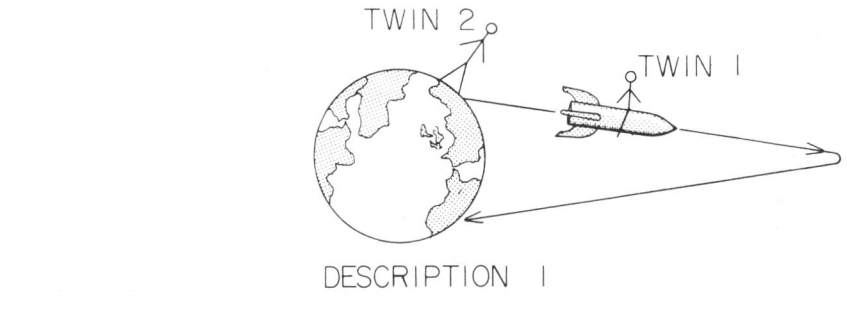

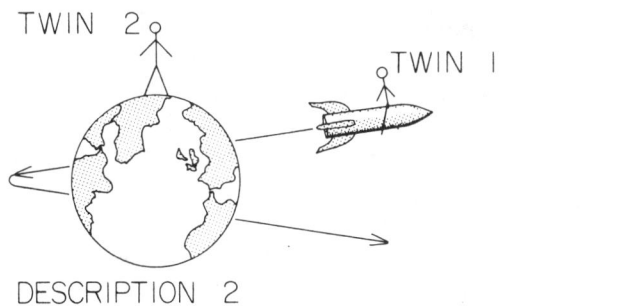

FIGURE 2. Clock paradox.

As has often been pointed out, this method of attempting to generate a paradox is defective for a very elementary reason. Such formulas as the time dilation equation apply only to clocks situated in inertial frames and not subject to accelerations. In order for the twins to move apart and then back together again, however, at least one of them has to experience accelerations.[7] Suppose that the second twin (who remains at A) experiences no accelerations. Then his brother has to accelerate to an appreciable fraction of the speed of light; moreover, when he reaches B he has to reverse his direction in order to travel back to A. This involves further acceleration. Which twin suffers the accelerations can be established empirically in an unambiguous way: crashing an automobile into a concrete wall at 200 miles per hour[8] is negligible compared with instantaneous reversal of direction when one is traveling at, say, 9/10 the speed of light. In the immortal words of John Cameron Swayze, the watch of the first twin really has to "take a licking and keep on ticking."[9] Only one of the twins need undergo what is known colloquially as a "bad trip." When accelerations are taken into account, any apparent symmetry in the situation vanishes, and the general theory can be invoked to ascertain unambiguously which of the two twins will be the younger.

While I do not deny the correctness of the general relativistic treatment of the clock paradox, I do agree with Adolf Grünbaum and others that it is somehow not completely satisfying intellectually.[10] It is, after all, the time dilation of special relativity that is suspected of spawning an inconsistency. Contradictions are not best treated by invoking a more complex theory, for it is a general and fundamental principle of logic that contradictions in a set of premises can never be eradicated by adding new premises; the only way to get rid of a contradiction is by removing some premises of the original set. If the special theory does contain an inconsistency we had better locate it, rather than covering it up with an augmented theory. Of course, it may be replied that the clock paradox holds no difficulties for special relativity because it cannot even be formulated in terms to which the restricted theory is applicable, and so the question of inconsistency cannot even arise. This answer does not seem fully adequate, however, for it is possible to formulate a version of the clock paradox which does not involve any accelerations. This version has been attributed to Lord Halsbury (see Figure 3).[11]

Assume we have three clocks, C, C', C". C is situated at rest at point A in its inertial frame K. K' and K" are the frames in which C' and C", respectively, are at rest. C' travels from left to right at constant velocity v relative to frame K. At the moment C and C' meet these two clocks are synchronized with each other by setting both to read zero. C' continues moving uniformly to the right; when it reaches a point B in K it meets a clock C" traveling in the opposite direction. At the moment of meeting C' and C" have the same reading, that is, they are locally synchronous at that point. C' continues moving to the right (and on out of the picture); C" moves inertially toward A where C has been situated for the whole time, and it passes C with a relative velocity $-v$. At the moment of meeting, the readings of C and C" are compared (see Figure 3). In this formulation we have dispensed with the twins, for there is no way to bring them back together again without subjecting them to accelerations, but we have made the relevant time comparisons. To be sure, we have not brought clock C' back to clock C, but we have brought the time reading of C' back via a third clock C" which was synchronized (locally) with C' as the two of them met at B. Local synchronization is taken as absolute and unproblematic; when two clocks are located at (approximately) the same place we can make a direct empirical comparison of their readings.

CLOCKS AND SIMULTANEITY IN SPECIAL RELATIVITY

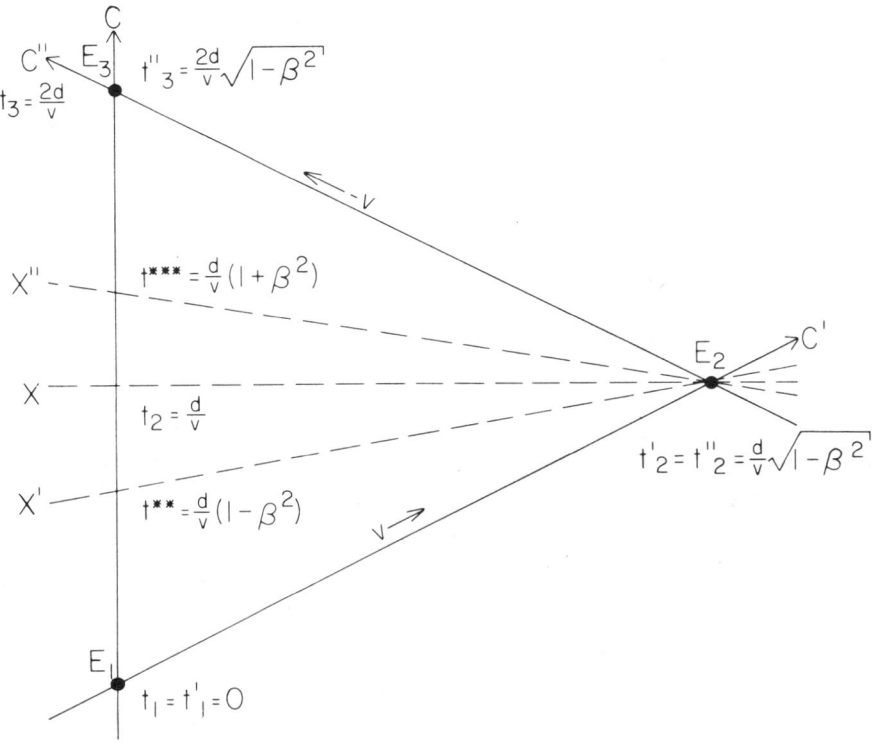

FIGURE 3. Clock paradox as seen from K (frame of clock C).

Since no accelerations are involved and all clocks are moving inertially, we can use the special theory to calculate the results. No considerations from general relativity need to be invoked.

Letting events E_1, E_2, and E_3 represent, respectively, the meeting of C with C' at A, the meeting of C' with C" at B, and the meeting of C" with C at A, we attempt to ascertain the times of these events according to the proper times of the various clocks. We let t stand for the time of clock C, t' the time of clock C', and t'' the time of clock C", and we let d be the distance between A and B with respect to frame K. To simplify our algebraic expressions, we follow the usual practice, introducing $\beta = v/c$ (or $-v/c$ indifferently since this quantity is always squared). We have set the clocks C and C' to read $t_1 = t'_1 = 0$ at E_1. With respect to frame K, C' travels a distance d at a velocity v, arriving at B (where it meets C") at $t_2 = d/v$. This is the time of event E_2 according to C. Applying the standard time dilation formula, we find that C' reads $t'_2 = (d/v)\sqrt{1 - \beta^2}$; this is the time of

E_2 according to C'. We have stipulated, moreover, that C" be set to agree with C' at E_2; hence, $t''_2 = t'_2$. Since, obviously, with respect to frame K, it takes C" just as long to get from B to A as it took C' to travel from A to B, C" will read $t''_3 = (2d/v)\sqrt{1 - \beta^2}$ at E_3, while C will show $2d/v$ as the combined time for the trips of C' and C" over distance d at speed v. Hence, C" is retarded relative to C when they meet. This outcome seems unambiguous when the analysis is carried out from the standpoint of frame K.

In spite of these straightforward results, one might still harbor a suspicion that the clock paradox has not been fully resolved, for even though we are using three clocks instead of two, the special theory of relativity says unequivocally that when clocks are in uniform motion with respect to one another *each* is retarded with respect to all the others. Until we have shown that this general fact is compatible with our analysis we have not completely handled the clock paradox.

In Figure 3, the broken lines X, X', X" represent the spatial axes in the frames K, K', and K", respectively, that is, they represent the simultaneity relations in each of the three inertial frames. Thus, in frame K, the clock reading of C which is simultaneous with E_2 is $t_2 = d/v$. With respect to frame K, clock C' runs more slowly than C, for C' reads $t'_2 = (d/v)\sqrt{1 - \beta^2}$ when C reads $t_2 = d/v$. C" also runs slower than C with respect to frame K for reasons of obvious symmetry. When we ask, from the standpoint of frame K', what clock reading on C is simultaneous with event E_2, the answer is different. According to K', $t^{**} = (d/v)(1 - \beta^2)$ is the reading on C when C' meets C", and that is less than $t'_2 = (d/v)\sqrt{1 - \beta^2}$; consequently, with respect to K', C runs slower than C'. Similarly, the clock reading on C which is simultaneous with E_2 with respect to K" is $t^{***} = (d/v)(1 + \beta^2)$. Again, from the standpoint of K", C is retarded with respect to C". The key to the whole problem lies in the discrepancy between the simultaneity relations of K' and K"; from the combined standpoints of K' during the interval from E_1 to E_2, and of K" during the interval from E_2 to E_3, the clock times of C between t^{**} and t^{***} simply drop out of the picture. The moments in that interval are not simultaneous (with respect to the simultaneity relations of K') with any part of the trip of C' from A to B, and they are not simultaneous (with respect to the simultaneity relations of K") with any part of the trip of C" from B to A.

In order to be quite sure that no paradox can be generated on the basis of Lord Halsbury's formulation, let us look at the whole situation from the standpoint of the frame K" in which C" is at rest (see Figure 4). From this viewpoint, there are two clocks, C and C',

CLOCKS AND SIMULTANEITY IN SPECIAL RELATIVITY

approaching C″ from the left at different velocities v and V. Using the composition of velocities formula, we find that $V = 2v/(1 + \beta^2)$. Before they arrive, the faster moving clock C′ catches up with C (this is event E_1) and moves on to meet C″ before C gets there. The meeting of C′ with C″ is E_2. Somewhat later, C arrives; this is event E_3. The lack of symmetry between the situation as seen from the standpoint of C and as seen from that of C″ becomes apparent by comparing Figures 3 and 4; they are in no sense reflections of each other.

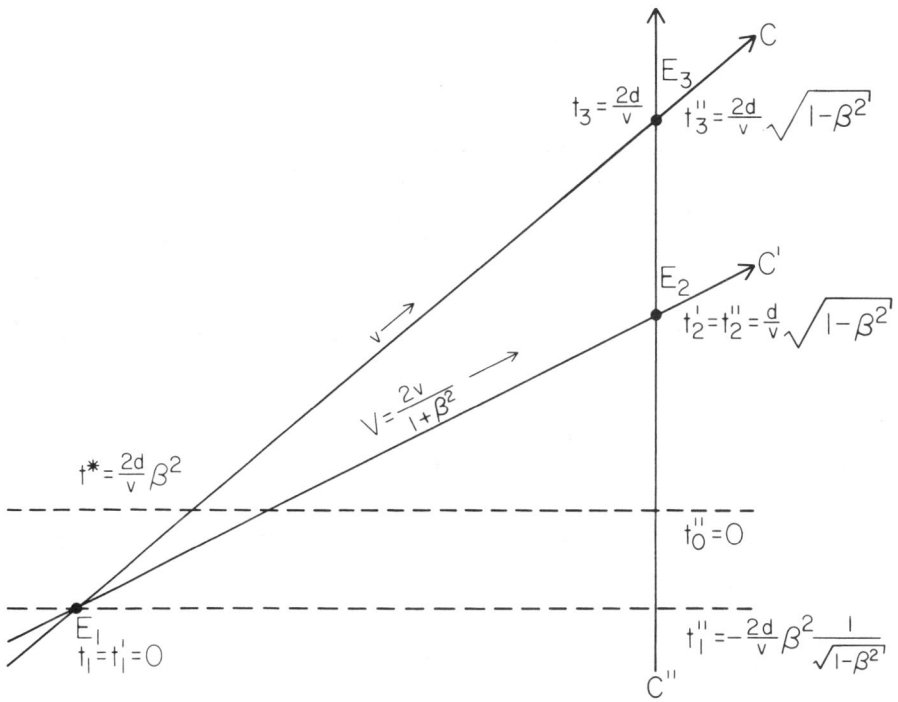

FIGURE 4. Clock paradox as seen from K″ (frame of clock C″).

The crucial point in analyzing the situation from the standpoint of K″ is recognition that E_1 (when C and C′ register $t_1 = t'_1 = 0$) is *not* simultaneous with the reading $t'' = 0$ on C″ according to the simultaneity relations of K″. That this must be the case is obvious from inspection of Figure 3, which shows the simultaneity relations of each of the three frames. Rather, from the standpoint of K″, the reading $t^* = (2d/v)\beta^2$ is simultaneous with $t''_0 = 0$. In K″, the time t''_1 of E_1 has the negative value $(-2d/v)\beta^2(1/\sqrt{1-\beta^2})$.[12] According to the simultaneity relations of K″, when C reads zero, C″ has a negative reading; when C″ reads zero, C has a positive reading. C is thus ahead of C″ on both of these occasions, and although C continues to run slower than C″ until they meet, C is still ahead of C″ at E_3 because C″ does not succeed in catching up with C before their meeting. From the standpoint of clock C″, it turns out, we get the same result as we got in analyzing the situation from the standpoint of C. On the occasion of their meeting (event E_3) clock C″ is retarded with respect to clock C. Precisely the same outcome can be calculated from the standpoint of K′.[13]

The foregoing resolution of the clock paradox obviously depends heavily upon appeal to the relativity of simultaneity. Indeed, the concept of simultaneity constitutes the key to the entire special theory of relativity. The well-known length contraction and time dilation effects rest directly upon the relativity of simultaneity, as we saw in the preceding chapter. At the age of sixteen, Einstein remarks in his "Autobiographical Notes," he had *wondered* about the role of the speed of light as a constant in Maxwell's electromagnetic theory, but it was only some years later that he saw how to *handle* the problem through a radical reanalysis of the concept of simultaneity.[14] The crucial character of Einstein's treatment of simultaneity has, of course, been widely recognized, but the fact that the revolutionary new analysis was a two-stage affair has not always been clearly noted. Einstein, however, treated the two stages separately and he explicitly acknowledged the distinction between them.

Einstein's first discussion of simultaneity in his famous 1905 paper, "On the Electrodynamics of Moving Bodies" (the original paper in which he first sets out the special theory), relates to the problem of establishing simultaneity relations within a single frame of reference. This is the problem of synchronizing clocks that are at rest with respect to one another in any inertial reference frame. This problem must be treated before moving on to consideration of the relativistic effects that arise from relative motion between two or more inertial frames, and Einstein discusses it in §1 of his original paper. His resolution of this problem is deceptively simple. Given two clocks situated at widely separated points A and B in an inertial frame, "we establish *by definition* that the 'time' required by light to travel from A to B equals the 'time' it requires to travel from B to A."[1] That Einstein meant to emphasize the definitional character of such synchrony is evidenced by the fact that the italics in the quoted passage are his, and by the fact that the title of §1 is "Definition of Simultaneity."

If a light signal sent from A is reflected at B so that it returns to A, a clock at A can be used to ascertain the total time required by light for the round trip from A to B and back again. Letting t_A be the time at which the light signal is sent from A, and letting t'_A be the time at which the light signal returns to A, Einstein goes on to say, "In agreement with *experience* we *further* assume the quantity

$$\frac{2AB}{t'_A - t_A} = c$$

to be a universal constant—the velocity of light in empty space" (my italics).[16]

Einstein thus enunciates two distinct principles regarding the speed of light in the first section of his 1905 paper. The second of these principles is an *empirical* hypothesis stating that the average speed of light on a round trip in a vacuum is a constant c. This principle, Einstein claims, is supported by experience; if true, it describes a fact of nature. The first principle equates the speed of light on each of the two legs of a round trip in a vacuum. Unlike the second principle, it is a *definition*. There can be no question of its truth or falsity, and under no circumstances can it properly be construed as articulating any fact of nature. The two principles can be contrasted and summarized as follows:

1. One-way light principle (a convention): On any round trip *in vacuo* the speed of light in the outbound direction is equal to its speed on the return trip.
2. Two-way light principle (a factual hypothesis): On any round trip *in vacuo* the average speed of light for the entire trip is equal to a constant c.

In conjunction, of course, these two principles imply that the speed of light on any one-way trip *in vacuo* is equal to the constant c. In most discussions of relativity it is the combination of the two principles that goes under some such heading as "the principle of constancy of light velocity," and it is this combined principle that we want to use in most contexts. Einstein, himself, uses the combined principle in subsequent portions of the 1905 paper. When we are trying to be careful about logical foundations, however, as was Einstein at the beginning of his famous paper, it is crucially important to distinguish the conventional from the factual components of the combined principle.

In order to underscore the distinction between the two light principles, let us consider a classic method for measuring the speed of light. This method is due to Armand Fizeau, who devised a cog-wheel arrangement to make the measurement (see Figure 5). A wheel with gaps between its teeth is made to rotate in front of a light source, which is sending a beam toward a mirror. When a tooth is in the light

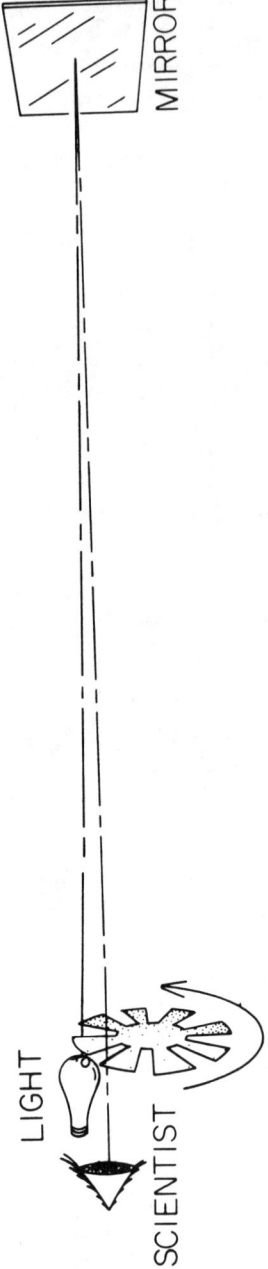

FIGURE 5. Fizeau's method for determining speed of light. At a suitable speed of rotation, the wheel will move the distance between two adjacent gaps in the time required for a light pulse to travel from the wheel to the mirror and back.

path no light gets through, but when a gap is there a light pulse passes through to the mirror, where it is then reflected back. Upon its return the reflected light pulse may encounter a tooth or a gap; if a gap is present it passes through once more, but if a tooth is present it will be blocked. By varying the angular velocity of the wheel and observing those velocities that allow the reflected pulse to go through, Fizeau was able to measure the speed of light. His method was similar in principle to that used later by Michelson; the measurement is patently the determination of a round-trip speed. It is worth noting that the Michelson-Morley experiment, which is often cited as evidence for the constancy of the speed of light, also compares average round-trip speeds over paths that are perpendicular to one another (see Figure 6). These are the kinds of experimental evidence that support the two-way light principle, but clearly, since they are methods of ascertaining and comparing round-trip speeds, they have no bearing whatever on the one-way light principle.

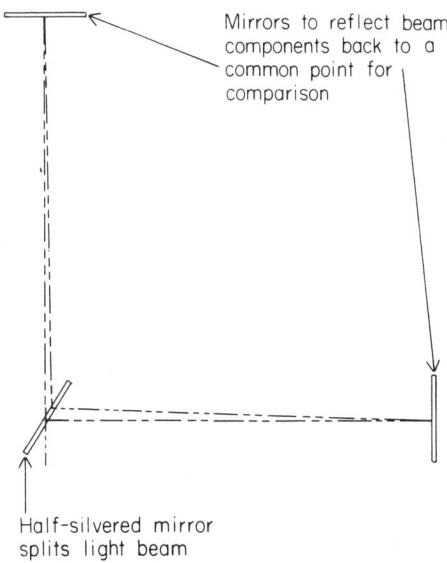

FIGURE 6. Michelson-Morley experiment (Schematic diagram).

Einstein could not reasonably have offered the one-way light principle as a definitional convention unless he held that the equality or inequality of the speed of light in the two opposite directions is *not* a matter of fact. Such an attitude is hard to accept. If we believe that the round-trip speed is not a matter of convention, it is difficult to see how the one-way velocity can fail to be a matter of fact. We all realize, of course, that every scientific statement rests in some fashion upon a bunch of semantic conventions. Once the standard semantic conventions have been adopted, however, the average round-trip speed of light is (as we have just seen) a matter of empirical fact, amenable to experimental ascertainment; the one-way speed, in contrast, must still remain a matter of convention after the standard semantic conventions have been adopted if Einstein's definition of simultaneity is to make sense. Thus, we must consider with some care whether it is possible to measure the one-way velocity of light.

To measure the one-way velocity we need, in principle, a path along which the light is to travel with a clock located at each end. One of these clocks measures the time of departure of a light signal, and the other measures its arrival at the other end of the path. Such measurements can have a significant bearing upon the determination of a one-way velocity only if they tell the same time—that is, only if they are synchronized with each other. Otherwise, the difference between the readings of the two clocks cannot represent a genuine time interval used in the one-way trip. But since the synchronization of spatially separated clocks is precisely the question at issue, we must consider further how that can be accomplished. Two methods suggest themselves.

First, we might try synchronizing the two clocks by sending messages back and forth. A person situated at A could send a radio message to someone located at B saying that his clock at A now reads, say, 12:00. The person at B who receives the message knows, however, that it took some time for the message to travel from A to B; hence, when he receives the message he knows that the clock at A no longer reads 12:00. As a matter of fact, radio waves travel at the same speed as light rays, so to know how long after 12:00 the message arrived would be tantamount to knowing the one-way speed of light.

This would pose no serious problem with Newtonian mechanics. If we wish to use two spatially separated clocks to measure the one-way speed of light, we should simply use much faster signals to synchronize them. It would be analogous to ascertaining the one-way speed of sound. We can tell how fast sound waves travel in one direction by noting, for example, how long after the lightning flash one hears the associated thunder (assuming, of course, that we know how far away the lightning bolt struck). The velocity of light is so much greater than the velocity of sound that it is harmless to regard the light signal as being transmitted instantaneously (that is, at infinite velocity). Similarly, in ascertaining the one-way speed of light, the time required for transmission of a signal that travels fast enough could simply be ignored. The clocks could be synchronized with super-light signals and then used to measure the one-way speed of light.

Newtonian physics imposes no upper limit on the speed at which messages can be transmitted, for material particles can be accelerated to arbitrarily large velocities. You can simply write your message on a piece of paper, enclose it in a suitable container, and then impose sufficient force for enough time to accelerate it to the desired velocity. Whether, according to Newtonian theory, gravitational influences are propagated infinitely rapidly does not matter; it is sufficient that material particles can travel at arbitrarily large finite velocities. Classical physics thus contains, in principle, the resources to measure the one-way speed of light to any desired degree of accuracy.

The situation in the special theory of relativity is fundamentally different. The speed of light is a limiting velocity. Even if we are not yet in a position to assign a numerical value to the one-way speed of light, we can say that it constitutes a speed limit for the transmission of signals. No signal sent from A at the same time as a light signal will reach B earlier than the light signal (sent directly *in vacuo*) does. This is what Reichenbach meant by calling light a "first signal." The speed of light is an upper limit for the propagation of any causal process, according to the special theory, and only causal processes can be used to transmit messages (and to synchronize clocks).[17] We need not worry about "tachyons" (particles which travel faster than light) at present, for, although they are the subject of interesting speculation, all attempts so far to discover them empirically have failed.[18] I shall return below to the question of the bearing it would have on special relativity if the existence of tachyons were established.

Reichenbach summarized the problem of synchronizing spatially distant clocks by means of signals in the following succinct terms:

> To determine the simultaneity of distant events we need to know a velocity, and to measure a velocity we require knowledge of the simultaneity of distant events. The occurrence of this circularity proves that simultaneity is not a matter of knowledge, but of a coordinative *definition*, since the logical circle shows that a knowledge of simultaneity is impossible in principle.[19]

Reichenbach is evidently attempting to reconstruct the rationale for Einstein's introduction of a stipulation regarding the one-way speed of light as a basis for a *definition* of simultaneity. We may begin by sending a light signal from A to B and, by reflection, back to A, and we can measure the time for the round trip on our clock at A. We may then announce the conventional decision that half of that time was taken by the light in going from A to B, and the remaining half was consumed in the return trip. The total round-trip time can be communicated to our colleague at B, and we can tell him of our *decision* to regard half of that as the time taken for the signal to go from A to B. The next time we send him a message telling him what the clock at A reads he can, without difficulty, synchronize the clock at B with ours. This is known as "standard signal synchrony."

Reichenbach introduced a useful notation for discussing syncrhonization of clocks by means of light signals.[20] If t_1 is the time at which a light signal departs from A traveling toward B, and t_3 is the time at which that signal, reflected at B, returns to A, then the time at which the signal arrived at B can be represented as

$$t_2 = t_1 + \epsilon (t_3 - t_1).$$

In order to secure the fact that the signal arrived at B after it was sent from A and before it returned to A, he imposed the condition

$$0 < \epsilon < 1.$$

Einstein's definition of simultaneity results when

$$\epsilon = \tfrac{1}{2}.$$

Reichenbach insists that this choice of ϵ results in far simpler description of the physical world than could be achieved with other choices, but the choice is motivated by convenience. It is not demanded by the "fact" that light travels at the same speed in the two directions of a round-trip, for *there is no such fact*. Other choices of ϵ would result in equivalent descriptions of the same physical facts. If someone were to make an alternative choice, Reichenbach claims, there would be no experimental way of proving him wrong by virtue of his commitment to the consequence that light travels at different speeds in the two opposite directions. This is the "cash value" of the claim that the one-way light principle is a convention.

While Fizeau's method of measuring the speed of light is, as mentioned above, applicable only to the round-trip speed, another historically important method of ascertaining the speed of light seems to give the one-way velocity we have been looking for. This method was involved when Olaf Römer accounted for periodic discrepancies in the eclipsing of the moons of Jupiter on the basis of the finitude of the velocity of light. When the earth and Jupiter are in opposition, light from Jupiter has to travel a path that is longer (by the diameter of the orbit of the earth) to reach us than it does when the two planets are in conjunction (see Figure 7). Knowing the diameter of the earth's orbit and using the apparent delay in the eclipsing of Jupiter's moons, Römer was able to make a rough determination of the speed of light on a *one-way* trip across the earth's orbit.

Römer's measurement of the one-way speed of light requires consideration of a second method (other than light signals) of attempting to synchronize clocks located at A and B. We begin with two clocks located at A and synchronized locally with one another. One of these clocks is then transported to B, where it is used subsequently to indicate the time. The most obvious difficulty with this method lies in the prediction by special relativity of the fact that transported clocks are affected in a way that depends upon the path and speed of transport. The relevant part of this consequence of the special theory can be checked without making any assumptions or stipulations about distant simultaneity. We can simply observe (as we already have in looking at the clock paradox) that the clock which was previously synchronized locally with the clock at A and then transported to B will, if returned to A, no longer be synchronized with the clock that remained at A. It would, consequently, be an entirely unwarranted assumption to suppose that the clock transported from A to B retains its synchrony with the clock at A.

It is easily seen that Römer's method for measuring the one-way velocity of light makes use, in effect, of a clock that is transported from one end of the one-way path to the other. In the half-year (approximately) between the approximate conjunction and the approximate opposition of the earth and Jupiter, the clock we use to establish the time of apparent eclipse of a moon of Jupiter travels with the earth from one end to the other of a diameter of its orbit. Unless the clock that travels with the earth is assumed to remain in synchrony with two *hypothetical* clocks that remain fixed (relative to the frame of the solar system) at the two ends of that diameter, the apparent discrepancy in the times of eclipsing of Jupiter's moons would not constitute a measure of the one-way velocity of light across the earth's orbit.[21] Hence, to construe Römer's method as a way of measuring the one-way velocity of light requires assumptions about clock synchrony that we are not yet entitled to make.

The standard Lorentz transformation equations of special relativity imply that the retardation R to which a clock traveling a distance d at velocity v is subject is given by

$$R = \tfrac{d}{v}(1 - \sqrt{1 - \beta^2}).$$

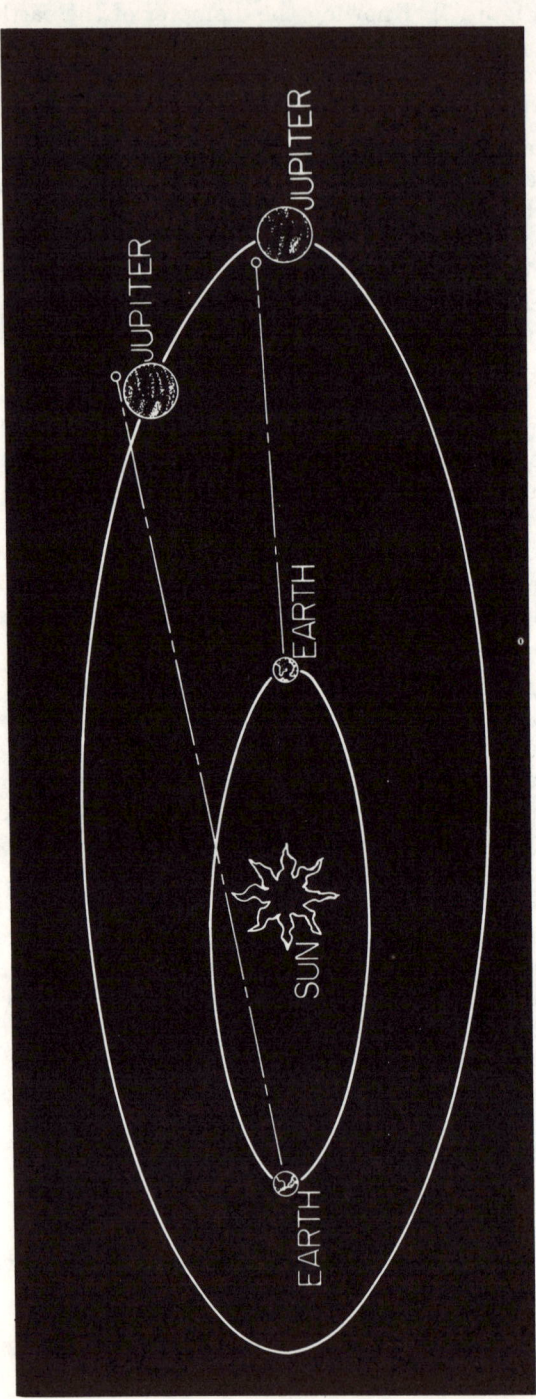

FIGURE 7. Römer's method for ascertaining light velocity. Because of large size of Jupiter's orbit, Jupiter moves through only a small part of it during half of an earth year.

Athough the expression on the right has v in its denominator, it is quite easy to show (using l'Hospital's rule) that R goes to zero as v approaches zero. Thus, the retardation of a clock we wish to send any distance d can be made as small as we like by choosing a sufficiently slow mode of transport. In order to synchronize two clocks located at different places, we need only take a third clock, place it beside the first clock, making sure that these two clocks are synchronized; then the third clock is carried *slowly* to the place of the second clock, which is then reset so as to agree with the third clock. Thus, the first two clocks are synchronized. It has been noted, for example, that the orbital velocity of the earth, which is of the order of 20 mi/sec, is a very tiny fraction of the speed of light, and consequently, Römer's earthbound clock would suffer negligible retardation in its semi-annual trip from one side of the orbit to the other.

There is, however, a basic logical problem that must somehow be handled. The term 'v' in the foregoing equation designates a one-way velocity, but until we have clocks synchronized with one another at the two ends of the path, no meaning can be assigned to *any* one-way velocity. It is not just the one-way speed of light that is in question. P. W. Bridgman, one of the authors who recognized the potentiality of slowly transported clocks for the establishment of synchrony relations, dealt with this problem by using the "self-measured velocity" of the clock.[22] Let d be the distance between A and B as measured in the frame in which the clocks at A and B are at rest. If we send a clock C' from A to B, it will show a certain time t'_1 at its departure from A and another time t'_2 upon arrival at B. The self-measured velocity at which C' traveled from A to B is $d/(t'_2 - t'_1)$. This is a sort of bastard velocity, for it involves a spatial interval as measured in the rest frame of the clocks at A and B divided by a time interval as measured in the rest frame of C'. Nevertheless, as Bridgman shows, this self-measured "velocity" can be used to define convergence to the limiting transport velocity of zero. In other words, we can substitute the self-measured velocity for v in the retardation formula, and we can then allow convergence of v to zero as a criterion of convergence of R to zero. It can be shown that the "slow clock transport synchrony" that is established in this way coincides exactly with Einstein's standard signal synchrony. Bridgman did not view this fact as altering the conventional status of distant simultaneity; rather, he merely regarded slow transport as providing an alternative operational *definition* of synchrony.

Brian Ellis and Peter Bowman, in a widely discussed paper, took the opposite point of view. They argued that the facts of slow clock transport render the concept of distant simultaneity non-conventional (except in a trivial sense).[23] In order to handle the problem of defining the transport velocity of the moving clock, they introduce an "intervening velocity" which is quite different from Bridgman's self-measured velocity, but it does the same job. The Ellis-Bowman slow clock transport synchrony coincides exactly with Bridgman's, and both coincide with Einstein's standard signal synchrony. Ellis and Bowman, in contrast to Bridgman, regard it as a physical *fact* that clocks synchronized by slow clock transport show the same time. They conclude that spatially separated clocks that have been synchronized by slow clock transport can be used to determine objectively whether light travels at equal speed on the two legs of a round trip. If the special theory of relativity is correct, then the speed of light is the same in the two directions, for that is the theory from which the coincidence of Einstein's standard signal synchrony (which stipulates the equality of the speeds in the two directions) with the Ellis-Bowman slow clock transport synchrony is deduced. They further maintain that Römer's method of ascertaining the one-way speed of light is free from any non-trivial conventional elements, and that consequently it provides an objective, factual determination of the one-way speed of light.

The Ellis-Bowman paper elicited a response consisting of papers by Grünbaum, van Fraassen, and myself, in which we tried to show that the method of slow clock transport, while entirely suitable for the purpose of establishing relations of distant simultaneity, is just as infected with non-trivial conventions as is Einstein's standard signal synchrony.[24] There is, I believe, no need to rehearse these arguments here. In my contribution to the aforementioned group of papers,[25] which has come to be known informally as the "Pittsburgh Panel," I enunciate a criterion for the non-triviality of conventions,[26] and by applying it to synchrony by slow clock transport, show that this method involves conventions which are as non-trivial as those involved in standard signal synchrony. I attempt to show, moreover, that this conventional element in clock transport synchrony does not depend upon the fact that synchrony can be destroyed by relative motion of clocks. Even if, contrary to fact, the clock which is transported from A to B and back to A were always in agreement with the clock which remained at A, clock transport synchrony would still involve the same kind of conventionality. Consequently, the fact that the retardation can be made arbitrarily small by transporting clocks slowly enough has no bearing upon the conventionality of distant simultaneity. The crucial basis for the conventionality of simultaneity in special relativity is not the time dilation phenomenon, but rather, the limiting character of the speed of light. This feature of special relativity constitutes one of its most fundamental departures from classical mechanics. If distant simultaneity is conventional, it is so because of a pervasive fact about the physical world, namely, that light is a first signal.

There have been many ingenious proposals for measurement of the one-way velocity of light. Reichenbach discusses several in his classical work, *The Philosophy of Space and Time*,[27] and others are still being advanced.[28] As one example, consider the familiar "light clock" which was used in Chapter 3 to show how the constancy of the round-trip speed of light implies time dilation (see Figure 8). This device consists of two mirrors mounted a fixed distance apart, perpendicular to some sort of support joining their midpoints, with a photon of light which is reflected back and forth between them. A unit of time can be defined as half of the interval required by the photon to make the round trip from one mirror to the other and back again. If we have such a clock in our frame K, and compare it with another such clock located in a frame K' which is in motion with respect to K, its unit interval (as seen from our frame) must be longer than ours because the path of its photon is longer than the path in our clock, and both photons travel at the same average round-trip speed with respect to any inertial frame of reference. Hence, the time of the moving clock is dilated.

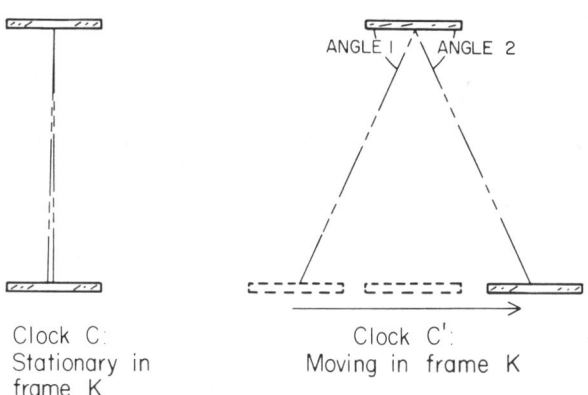

Clock C:
Stationary in
frame K

Clock C':
Moving in frame K

FIGURE 8. Light clock.

It has been suggested that we can establish the equality of the two one-way speeds by examining the path of the photon in the moving clock C' from the standpoint of our frame of reference. If angle 1 (the angle of incidence) is equal to angle 2 (the angle of reflection) then the photon is traveling at the same speed before and after the reflection from the upper mirror. This argument is inconclusive, however, for it contains a tacit synchrony assumption. In order for the equality of angles 1 and 2 to entail the equality of the speeds in the two directions, it is necessary for the support bar of clock C′ to be oriented perpendicular to the line of motion of C′ in reference frame K. This in turn requires that points A and B in our frame coincide with points A′ and B′ in the moving frame *at the same time*. Thus, the use of this method to ascertain empirically the equality of the two one-way speeds is vitiated by the need to have already established the simultaneity of events occurring at A and B, the endpoints of the one-way path.

If the foregoing argument about the behavior of the light in the moving clock were adequate to establish the equality of the two one-way speeds, then a similar argument based upon the well-known phenomenon of aberration of starlight would be equally adequate. Suppose we have a star that lies in a line perpendicular to the plane of the earth's orbit (the ecliptic) at a given point in the earth's orbit, and assume that it is distant enough to render negligible any discrepancy from perpendicularity at any point of the orbit. Assume, that is, that the star would appear in a line normal to the plane of the earth's orbit by an observer at rest in the frame of the fixed stars at any point in that orbit. Such a star would, of course, have no measurable parallax. The phenomenon of aberration of starlight consists in the fact that to view such a star from a telescope moving with the earth relative to the fixed stars, the telescope must be deflected slightly from the perpendicular orientation in order to compensate for the finite time interval required by the light to traverse the length of the telescope (see Figure 9). Knowing the length of the telescope, and knowing the speed of the earth relative to the fixed stars, we could calculate the one-way speed of light as it passes through the telescope if we could ascertain what orientation of the moving telescope is absolutely parallel to the path of the light ray in the rest frame of the fixed stars. But establishing parallelism or perpendicularity of a moving telescope (or light clock) begs the question of simultaneity of events located at opposite ends of the one-way path, and consequently fails to provide a basis for empirical determination of the one-way velocity of light. It is worth noting that the difficulty in using the light clock or the aberration of starlight as a means of ascertaining the one-way speed of light is identical to that involved in the use of such "unreal sequences" (discussed by Reichenbach) as the motion of the point of intersection of two rulers in motion with respect to one another.[29] It is also interesting to note that, according to Shankland, Einstein had been studying experiments on aberration of starlight very intensively at the time just before he composed the famous 1905 paper on special relativity.[30]

CLOCKS AND SIMULTANEITY IN SPECIAL RELATIVITY 113

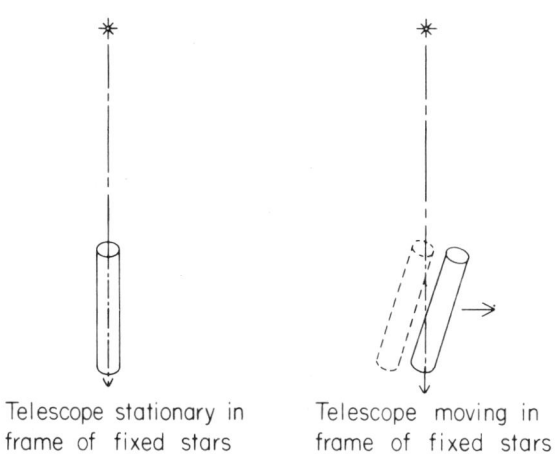

Telescope stationary in frame of fixed stars

Telescope moving in frame of fixed stars

FIGURE 9.   Aberration of starlight.

I do not wish to assert dogmatically that no method can be devised for genuinely circumventing the need for a non-trivial convention such as Einstein employed regarding the one-way speed of light, for that would amount to a claim of ability to foresee the results of all future experiments. At present, I do not know of any such method. It seems to me, therefore, that to the best of our present knowledge, the equality or inequality of the speed of light in the two opposite directions of a round trip must be regarded, not as a matter of fact, but rather as a matter of convention. Our choice of a value of ϵ seems, therefore, to characterize the language we choose for our description of the world; it does not seem to characterize any non-linguistic feature of the physical world itself.

Given the difficulty of the problems associated with Einstein's definition of simultaneity, one might wonder why a great many textbook introductions to special relativity make no mention of the need for any such definition. Such presentations frequently mention the constancy of the speed of light, but they fail to make any reference to the distinction between the one-way light principle and the two-way light principle. Nothing is said about methods of measuring the one-way velocity of light. It is hard to believe that authors of such books want to maintain that Fizeau's experiment or the Michelson-Morley experiment establish the constancy of the one-way speed of light when manifestly they do no such thing. There is, I believe, a different explanation.

If one is anxious to deal with the interesting parts of the special theory, it is tempting to get busy and derive the Lorentz transformation equations for frames which are in uniform motion with respect to one another. A frequent approach is to introduce, casually and without fanfare, what is known as the "reciprocity condition," namely, if frame K' is moving at velocity v with respect to frame K, then frame K is moving at velocity $-v$ with respect to frame K'. As a matter of fact, this is precisely the approach used in the preceding chapter. This assumption seems so natural that it goes by almost unnoticed; nevertheless, it turns out that the one-way light principle can be deduced from it.[31] Thus, in the context of special relativity, the reciprocity condition is tantamount to Einstein's stipulation concerning the one-way speed of light. Once the reciprocity condition is introduced, the Lorentz equations can be derived, but they are derived on the basis of the assumption that the speed of light is equal in the two legs of the round trip. This stipulation can be expressed in Reichenbach's notation by choosing 1/2 for the value of ϵ. Hence, the Lorentz equations in their standard form embody the convention, $\epsilon = 1/2$. The convention has been smuggled in quite surreptitiously via the reciprocity condition. The reciprocity condition is, consequently, a good deal more significant than it may appear at first blush.

The thesis that the value of ϵ may be freely chosen within wide limits is known as the *conventionality of simultaneity*. Once the normal choice $\epsilon = 1/2$ has been made, and we adopt Einstein's standard signal synchrony in all inertial frames, it is easy to show that spatially separated events which are simultaneous with respect to one inertial frame are not simultaneous with respect to any other inertial frame that is in motion relative to the first frame. This result is known as the *relativity of simultaneity*. The conventionality and relativity of simultaneity constitute the above-mentioned two stages in Einstein's revolutionary analysis of simultaneity.

The relativity of simultaneity rests upon the conventionality of simultaneity in the following sense. Two events e_1 and e_2 which are simultaneous in frame K on the basis of standard signal synchrony will not be simultaneous in K' (where K' is in motion relative to K) *according to standard signal synchrony*. By a judicious selection of value other than 1/2 for ϵ in frame K', however, it is possible to set up non-standard synchrony relations which will render e_1 and e_2 simultaneous in K' as well as in K. The conventionality of simultaneity makes it possible to erase the relativity of simultaneity.[32] The relativity of simultaneity thus embodies the standard convention, $\epsilon = 1/2$.

Similar remarks apply to length contraction and time dilation. Since the length of a moving object is given by its simultaneity projection in the stationary frame, a change in the synchrony relations entails a change in the length contraction relations. Indeed, it can be shown that, by judicious selection of a non-standard value of ϵ in frame K, the length of an object in motion in frame K can turn out to be equal to its own rest length. Furthermore, if we have a clock C′ moving from A to B in frame K, which is locally synchronized with the clock at A when it departs, it will be retarded with respect to a clock at rest at B upon its arrival there, *if the clock at B is in standard signal synchrony* with the clock at A. Again, however, by judicious choice of a value of ϵ other than 1/2, it is possible to synchronize the clocks at rest at A and B in K in such a way as to yield the result that C′ is locally synchronous with the clock at rest at B when it arrives there. The length contraction and time dilation relations thus involve the standard choice of ϵ, and within limits they can be erased by adoption of suitable non-standard synchrony relations.

It might be tempting, at this point, to wonder whether the special theory of relativity has any distinctive content at all, or whether, rather, it involves nothing more than a definition of simultaneity which departs from that used in classical physics. The answer to this query is already available: as we have seen, special relativity has certain testable consequences which differ from the consequences of classical physics (for example, the observed retardation in Lord Halsbury's three-clock setup; the muon decay phenomena; the length needed for the Stanford Linear Accelerator). Special relativity in its standard form has factual content, but it also has results that express conventional choices. To sort out these two kinds of elements can be a tricky job, but it is one that has to be done if we are to have a clear conception of the logical foundations of special relativity.

Both Einstein and Reichenbach maintain that the choice of ϵ is a matter of convention, and that one could, without contradicting any empirical fact, make a different choice. The price of a non-standard choice is simply a great deal of added complexity in the statement of the theory, without any change of content. They then proceed to invoke the usual convention and set ϵ equal to 1/2. From there they go on to derive the Lorentz transformations. The equations they derive are, therefore, infused with this conventional element.

In a most interesting and constructive response to the thesis of Ellis and Bowman, John Winnie has taken a different tack.[33] Instead of choosing a particular value for ϵ and deriving the Lorentz equations in their customary form, he deliberately foregoes any choice and proceeds to derive the equations of the special theory with ϵ present as a free variable. His rationale is quite straightforward. If the value of ϵ is a matter of convention, then it ought to be possible to state the "factual core" of special relativity without introducing the convention at all. To carry out this task he derives what he calls the "ϵ-Lorentz transformations." The burden he thus assumes is twofold: first, he must show how these transformation equations can be derived from premises that do not contain hidden one-way velocity assumptions, and, second, he must show that these equations yield the observational results for which the special theory must be held accountable.

Winnie adopts three postulates for special relativity, all of which are free from assumptions about one-way velocities.[34] They are:

1. *The round-trip light principle:* The average round-trip speed of any light-signal propagated (*in vacuo*) in a closed path is equal to a constant c in all inertial frames of reference.

This is simply our two-way light principle, generalized to hold for any closed path.

2. *Principle of equal passage times:* Let K and K' be two inertial frames in relative motion, and let A and A' be arbitrary points on the x-axes of K and K' respectively. Let Δt be the time interval in K of the passage of a rod at rest in K' of rest length s past the point A in K, and let $\Delta t'$ be the time interval in K' of the passage of a rod at rest in K of the rest length s past the point A' in K'. Then $\Delta t = \Delta t'$.

This principle is *not* equivalent to the reciprocity condition, for it does not imply that the velocity of K with respect to K' is equal to minus the velocity of K' with respect to K. Nor is it tantamount to the condition of reciprocity of relative lengths, for it does not imply that the lengths of the two rods whose rest lengths are both s in their respective frames have equal lengths as measured in the systems with respect to which they are in motion.[35] The principle of equal passage times entails only that the ratio of moving length to relative velocity is equal for the two rods. It is clear that this principle is free from one-way velocity assumptions, for only one clock in each inertial frame is involved. This principle is thus independent of all considerations relating to the synchronization of spatially separated clocks in any inertial system.

For reasons of expository convenience Winnie adopts a linearity principle, which says that any motion which is uniform straight-line motion in one inertial frame must likewise be uniform

CLOCKS AND SIMULTANEITY IN SPECIAL RELATIVITY   117

straight-line motion in any other inertial frame. This principle says, in effect, that Newton's first law holds equally in all inertial frames. But, inasmuch as the linearity condition is not obviously free from one-way velocity assumptions, Winnie remarks that it could be replaced by

> 3. *Principle of proportional passage times:* The passage time in K (or K') of a rod of rest-length L_0 is directly proportional to the rest length L_0.[36]

This principle, in conjunction with the other two, is sufficient to derive the linearity principle, and it is evidently free from one-way velocity assumptions. From these three principles Winnie derives his ϵ-Lorentz transformations.[37]

The main value in Winnie's approach lies in its ability to distinguish those consequences of the special theory of relativity which depend upon a particular choice of ϵ from those which are valid regardless of the choice. Thus, for example, we cannot uniquely assign a velocity to a muon traveling in a straight line relative to our frame of reference nor a time of travel without making a definite choice of ϵ. By applying the ϵ-Lorentz transformations we can, however, uniquely determine the place at which the average muon will decay if we know the average lifetime of a muon at rest in our frame of reference.[38] Observable facts which occur every day in particle accelerators can be derived from the factual core of the special theory without invoking any conventional choice of ϵ; consequently, the observations of such results are irrelevant to the question of the one-way speed of light, for they are compatible with any decision we choose to make.

For another example, consider our earlier discussion of the clock paradox in terms of Lord Halsbury's three-clock setup. In that context, we noted that the readings of adjacent clocks could be compared directly. Thus, we could ascertain the agreement of C with C' when they met (event E_1) and the agreement of C' with C" when they met (event E_2), and we could determine empirically whether C" was or was not retarded relative to C when they met (event E_3). The predicted retardation is an observable consequence which must occur, regardless of any conventions or stipulations, if special relativity is a true theory. This is a time dilation effect, and Winnie shows that its occurrence (and amount) is derivable from his ϵ-Lorentz transformations without introducing any one-way velocity assumptions.[39] This fact is another consequence of the "factual core" of special relativity.

When the standard Lorentz transformations (which embody the convention $\epsilon = 1/2$) are used, one can derive statements about clock readings that are not directly verifiable by observation. For example, referring back once more to our discussion of the clock paradox, we can ask what reading appears on clock C when (from the standpoint of Frame K) C reads d/v. We can derive the answer $(d/v)\sqrt{1 - \beta^2}$, but we cannot check the answer directly because the two clocks are not adjacent. If we had a clock at B that had previously been synchronized with C, we could compare its reading with that of C', but the result obviously depends upon the manner in which the clock at B was synchronized with the clock at A. Winnie's ϵ-Lorentz transformations do not provide unique answers to questions of this sort; the answers turn out to functions of ϵ. This illustrates the manner in which some questions have convention-laden answers, while others have convention-free answers that can be tested by direct observation.

A similar distinction can be made concerning length contraction phenomena. Using the standard Lorentz transformations we can compute the length of a moving rod, but, as we have already seen, this answer depends upon our choice of ϵ. A different choice of ϵ will yield a different length of the moving rod. This sort of contraction is, therefore, convention-laden. But there is no way to check such contraction phenomena empirically apart from performing a simultaneity projection of the moving rod, and that obviously involves commitment to some particular synchrony.

Einstein, however, proposed a "twin rod" experiment that gives rise to an observational result which is independent of any synchrony or one-way velocity assumptions. Suppose we have two rods of equal rest length moving in opposite directions in our frame of reference. Let the lefthand ends of the rods be A and A', respectively, and the righthand ends B and B', respectively. As the rods pass one another, there will be a point A* in our frame at which the two lefthand ends coincide, and another point B* at which the righthand ends coincide. We make no assumption about the one-way speeds at which the two rods are moving, and we make no commitment to whether the coincidence of the lefthand ends at A* is simultaneous with the coincidence of the righthand ends at B*. We merely mark these two points in our frame when each coincidence occurs, and we measure the distance between them at our leisure. Using the ϵ-Lorentz transformations it is possible to show that the distance A*B* < AB = A'B'. This is an observable contraction effect whose occurrence follows from the factual core of the special theory; consequently, its occurrence does not have any bearing upon the question of the one-way speed of light.[40]

Although Winnie does not show exhaustively that all observational consequences of special relativity are derivable by means of the ϵ-Lorentz transformations of the convention-free formulation, he does deal with enough of the typical problems to establish a strong presumption that this is true. It seems to me that the burden of proof now falls upon anyone who wishes to maintain the factual character of the one-way speed of light to produce an observable consequence of the theory which does not follow from the ϵ-Lorentz transformations. And it is strongly to be recommended that any alleged experimental demonstration of the equality of the two one-way speeds be analyzed in terms of the ϵ-Lorentz transformations, in order to guard against the unwitting introduction of the convention $\epsilon = 1/2$ via the standard Lorentz transformations. For example, it follows from the ϵ-Lorentz transformations that standard signal synchrony must coincide with slow clock transport synchrony. From this it follows that Römer's method does not constitute an independent method for ascertaining the one-way speed of light within the special theory. It shows that whatever value we assign to ϵ, slow clock transport synchrony must agree with standard signal synchrony. Römer's method does not constitute a measurement of the value of ϵ; instead, it constitutes a test of factual content of special relativity. If an experimental determination of the one-way speed of light by Römer's method were to establish it to be other than c, this would not be an experimental proof that $\epsilon \neq 1/2$; it would, instead, be an experimental disproof of the special theory of relativity. For Römer's method involves adoption of slow clock transport synchrony, and the ϵ-Lorentz transformations entail that this must coincide with $\epsilon = 1/2$.[41]

According to Newtonian mechanics, as I mentioned above, material particles can be accelerated to arbitrarily high velocities, well beyond the speed of light. If Newtonian mechanics were true, such particles could, in principle, be used to synchronize clocks and to establish relations of absolute simultaneity. Suppose a light signal is sent from A toward B at time t_1 according to a clock C at rest at A; upon reaching B it is reflected back, the time of its return being t_3 according to the same clock C (see Figure 10). If we assume that the speed of light is the same in both directions, we set the time of arrival of the light signal at B equal to $t_2 = t_1 + (1/2)(t_3 - t_1)$. In the context of Newtonian mechanics, this is a physical hypothesis that can be verified to any desired degree of accuracy through the use of super-light signals. At a time t^*_1, which is later than t_1 but earlier than t_2, a super-light signal could be sent from A which would arrive at B at the same time as the light signal sent to t_1. In addition, let the super-light particle be reflected back to A upon arrival at B—such "reflection" might consist simply in the triggering of an emission of another particle similar to the one which traveled from A to B. In any

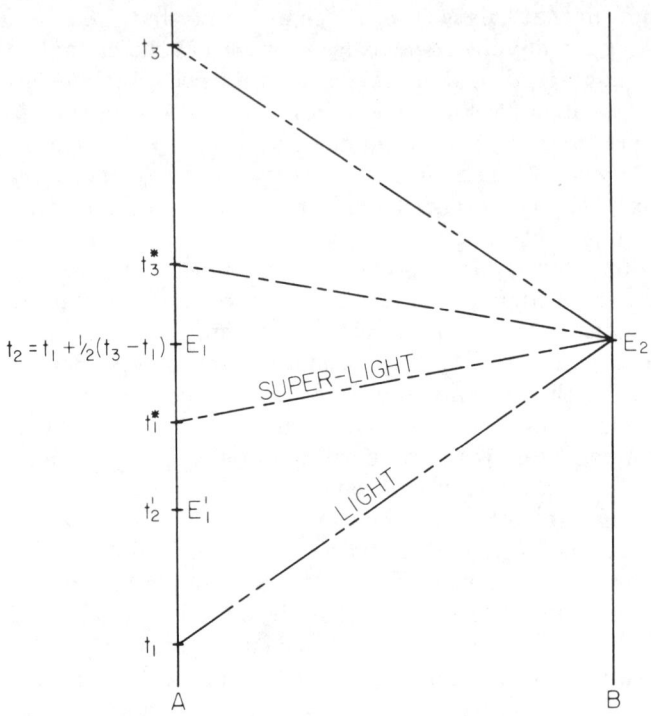

FIGURE 10. Ascertaining simultaneity with super-light signals.

case, the super-light signal which returns from B to A arrives at A at t^*_3, a time (on clock C) which is between t_2 and t_3. In view of these facts, we could say now that the time of the arrival of the light signal and the super-light signal must lie somewhere between t_1^* and t_3^*. This interval is smaller than that between t_1 and t_3; moreover, it can be made arbitrarily narrow by the use of arbitrarily fast super-light signals. We are not making any assumption about the relative one-way velocities of the fastest super-light signal that is employed; all we are relying upon is the assertion that the super-light signal arrived at B sometime between its emission from A and its return to A. If we were to say that our light signal arrived at B at a time t'_2 which is earlier than t^*_1, then we would be involved in asserting that our super-light signal (which arrived at B at the same time as the light signal) arrived at B before it was sent from A. This would contradict the causal features of our assumption that the super-light signal was sent *from* A *to* B, and that it is capable of transmitting information

CLOCKS AND SIMULTANEITY IN SPECIAL RELATIVITY   121

from A to B. Similar difficulties would obviously arise if we said that the light signal arrived at B after t^*_3.

Suppose that there are two events, E_1 and E_2, which occur at A and B, respectively; let each of these events be the radioactive decay of an unstable nucleus. E_1 occurs at the clock time t_2, while E_2 occurs at the moment at which the light signal and the super-light signal reach B. Suppose, moreover, that no matter how fast a super-light signal we employ, it must be sent from A before E_1 occurs if it is to reach B by the time E_2 occurs. Analogously, suppose that no matter how fast a super-light signal we use, it must be sent from B before E_2 if it is to reach A by the time E_1 occurs. E_1 is thus singled out as the unique event at A that is simultaneous with E_2. If, for example, an event E'_1 which occurred at A before E_1 were said to be simultaneous with E_2, we could refute the claim by pointing out that the event E'_1 is earlier than the sending of a super-light signal which arrives at B just as E_2 is occurring. Hence, E'_1 must be earlier than E_2 and cannot be simultaneous with it. By parallel reasoning, we can say that no event later than E_1 can be regarded as simultaneous with E_2.

The foregoing situation has been described from the standpoint of a particular frame of reference, as must be done if we want to assign space-time coordinates to the events we are discussing. But there are certain features of the situation which hold true in any reference frame we might choose—these are the *invariants*.[42] They are objective physical facts that do not depend upon our way of describing them. Regardless of the reference frame we choose, event E'_1 is causally connected with the emission of a super-light signal from A at t^*_1 (for example, by a causally connected series of time readings on the clock C), and this super-light signal constitutes a causal connection between its own departure from A and its arrival at B at the moment E_2 is occurring. Hence, E'_1 and E_2 cannot be considered simultaneous *in any reference frame whatever*, for the causal connection between them is an invariant which must be acknowledged to hold regardless of reference frame. If these same events, E'_1 and E_2, were observed from a different reference frame which is in motion relative to our frame, and if these two events were pronounced simultaneous on account of the time of their occurrence as registered on the clocks of that reference frame, we would simply have to say that something had gone wrong with the clocks in that frame and that they are no longer synchronized.

It is equally an invariant fact that the events E_1 and E_2 cannot be causally connected with one another; this is true regardless of reference frame, and, consequently, these two events must be considered simultaneous in every reference frame. Our friends in the moving reference frame must reset their clocks accordingly. We see, therefore, that in the Newtonian world of arbitrarily fast signals and causal processes, both the conventionality and the relativity of simultaneity are untenable. Arbitrarily fast signals yield absolute simultaneity of the strongest sort; the presence of the relativity of simultaneity in special relativity hinges crucially upon the existence of a finite upper speed limit on the propagation of causal processes and signals.

Although it has been recognized from the beginning that special relativity precludes the acceleration of particles from velocities less than that of light to velocities greater than c, it has also been observed that its equations (for example, for composition of velocities) do not rule out the possibility of particles that *always* travel at superlight velocities. Such particles could not, of course, be decelerated to speeds lower than c. In recent years there has been considerable speculation about the existence of such particles, called "tachyons," and some serious effort to detect them experimentally.[43] Although all such efforts to date seem to have failed (as mentioned above), it is still interesting to consider what the implications for special relativity would be if tachyons were to be discovered. We can say, at the very least, that their existence would have severe repercussions for the concept of causality.

Suppose we have a rocket ship that stands for some time at rest relative to a space station, but at a substantial distance away from it, and then it rapidly accelerates to a speed that is a very large fraction of c (see Figure 11). Immediately after achieving the desired velocity the rocket ship sends a tachyon message back to the space station. Immediately upon receipt of this message the space station sends a reply by tachyons that reaches the rocket ship sometime before it began its acceleration. In fact, the message from the space station to the rocket ship might trigger a detonator that destroys the rocket ship. Let E_3 be this explosion. Thus, the rocket ship is destroyed before it sends the message that initiated the process leading to its destruction. It seems to me that we are faced with the following dilemma: either tachyons cannot be used to send messages (and if not, why not?) or they can be used to establish absolute synchrony and absolute simultaneity, thereby eliminating the relativity of simultaneity which is so fundamental to the entire special theory.[44]

In special relativity, although the spatial distance between two distant events, as well as their temporal separation, are relative to a frame of reference, the space-time interval between them is an invariant. This interval, whose square is equal to the square of the temporal separation minus the square of the spatial separation, is a constant. It depends in no way upon the choice of the reference frame in which the spatial and temporal separations are measured. If the

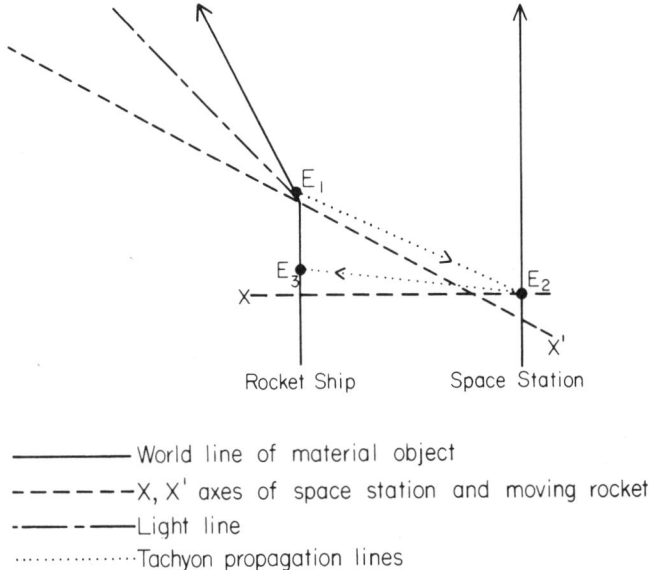

FIGURE 11. Causal anomaly with tachyon signal.

square of the space-time interval between two events is positive, the interval is said to be timelike (see Figure 12). The interval between E_0 and E_1 is timelike; this means that it is physically possible to have a material reference frame in which both E_0 and E_1 occur at the same place, though at different times. In any other material reference frame there is some time later than the occurrence of E_0 at which a light signal can be sent from the place at which E_0 occurred, and which will arrive at the place at which E_1 occurs at precisely the time of its occurrence. It is physically impossible to find a reference frame in which they occur at the same time—that is, simultaneously—but at different places. When the square of the interval between two events is zero, we say that they have a lightlike interval. The interval between E_0 and E_2 is lightlike. This means that it is physically possible for a light signal that originates in the event E_0 to travel directly to the space-time point at which E_2 occurs. If the square of the interval is negative the interval is said to be spacelike. E_0 and E_3 have a spacelike separation. This means that (according to standard signal synchrony or any other permissible synchrony definition) there is a reference frame in which E_0 and E_3 are simultaneous. It means, moreover, that if one wanted to send a light signal from the place at which E_0 occurs (with respect to any possible material reference

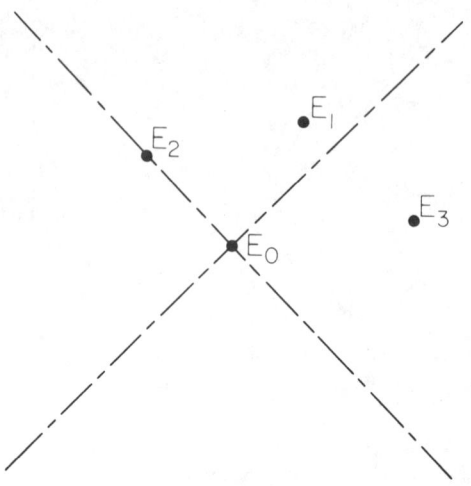

FIGURE 12. Types of space-time interval.

frame) which will arrive at the space-time point at which E_3 occurs, it is necessary to send it at a time earlier than the occurrence of E_0.[45]

This is one natural way of interpreting the sign of the square of the interval. A zero interval squared means that two events are directly connectable via a first signal; a positive interval squared indicates that direct connectability via a first signal lies in the future; a negative interval squared indicates that connectability via a first signal lies in the past. If, however, light were not a first signal, there would hardly seem to be any physical justification for according light such a significance in our space-time schemas. If there were another first signal, faster than light but still propagated at a finite velocity, it could, of course, replace light in the foregoing discussions. But if, as in Newtonian theory or in tachyon theory, there were no first signal with a finite velocity, it would seem that the physical foundation for the distinction between timelike and spacelike separations would vanish. Thus, I should think, a convincing experimental demonstration of the existence of tachyons would demand severe revisions in the foundations of the special theory of relativity, perhaps even to the extent of eliminating the conventionality and relativity of simultaneity altogether.

Although Reichenbach was clearly aware of the distinction between conventionality of simultaneity with respect to a single inertial reference frame and relativity of simultaneity between two or more reference frames in motion with respect to one another, he was not particularly scrupulous in his use of such terms as "convention," "definition," and "relativity." The equivocal use of such terms does create a danger of confusion, and the more careful usage of subsequent authors (for example, Grünbaum and Winnie) helps to lessen the risk. It also suggests that we look carefully at the concepts of conventionality of simultaneity and relativity of simultaneity to ascertain the relations between them. We have seen that in Newtonian mechanics simultaneity is neither (non-trivially) conventional nor relative, and we have been led to suspect that the same situation would obtain if there were, in fact, such things as tachyons. We have also seen that the Einstein-Reichenbach interpretation of special relativity renders simultaneity both conventional and relative. We have seen, moreover, that the conventionality of simultaneity can be exploited to eliminate some types of relativity; indeed, it can easily be used to eliminate the relativity of simultaneity entirely. By a conventional decision we could designate some particular inertial system the privileged frame of reference, and by another conventional decision we could choose standard signal synchrony *in that frame* as determining the relations of simultaneity that obtain *non-relatively* among all events in the entire universe. Such conventional choices would be contrary to the spirit of the special theory of relativity in some sense; in addition, they would prove inconvenient and inefficient. I cannot see, however, that they would lead to any refutable claims (if special relativity is true). On Reichenbach's construal of special relativity, such a set of conventions would lead simply to another description of the physical world which is factually equivalent to the special theory in its standard formulation.[46]

These considerations lead us naturally to wonder about the fourth possibility, namely, a simultaneity relation which is non-conventional but still relative. This is a possibility that, I believe, should be considered seriously. It is the result to which one is led by accepting the Ellis-Bowman thesis of non-conventionality of simultaneity by slow clock transport. Quite clearly, the simultaneity relation established by slow clock transport is relative to a particular inertial reference frame, for slow clock transport synchrony coincides with standard signal synchrony. It is well known (as we saw in chapter 3) that standard signal synchrony leads to relativity of simultaneity; two events that are simultaneous with respect to one frame of reference will be non-simultaneous in a different reference frame. The fact that simultaneity becomes relative to a reference frame does not, however, entail that the relation is non-trivially conventional. It means, merely, that a relation that had been considered binary in Newtonian mechanics is ternary in special relativity. Ternary relations are just as capable of being objective as are binary relations. We have maintained, for instance, that one event can stand in the relation of *earlier than* to another event objectively and invariantly when a suitable sort of causal relation connects them. Similarly, under easily specifiable causal relations, we are entitled to say objectively and invariantly that one event is *between* two others in time. No non-trivial conventions need to be invoked for either type of statement.

Similarly, one might argue, it is possible for the statement, "Event E_1 is simultaneous with event E_2 with respect to reference frame K," to represent an objective state of affairs that involves no use of non-trivial conventions. The fact that the statement, "Event E_1 is not simultaneous with event E_2 with respect to reference frame K'," may be true is obviously irrelevant to the question of non-trivial conventional component. One could, for instance, argue that simultaneity relative to a given frame is an objective fact of nature by maintaining that it is an objective fact that the one-way speed of light is c in every reference frame. To maintain that the one-way speed of light is a constant in every reference frame (an invariant of the theory) one needs to provide an objective method for ascertaining the one-way speed (without invoking non-trivial conventions such as are contained in Einstein's standard signal synchrony). Ellis and Bowman believe they have provided such a method based on slow clock transport synchrony, and that consequently, Römer's method of ascertaining the one-way speed of light does exactly the desired job. Thus, they claim, the one-way speed of light is an invariant fact of nature, and this implies that the ternary relation of simultaneity relative to an inertial frame also represents an invariant factual relation.

I do not believe that Ellis and Bowman provided a satisfactory argument to establish the relativity and non-conventionality of the relation of simultaneity. The most that has been shown, and perhaps the most that can be shown, is that it is an objective and invariant fact that standard signal synchrony coincides with slow clock transport synchrony (and that both of these coincide with another type of clock transport synchrony which does not require slow transport[47]). The coincidence among these various types of synchrony entails neither the conventionality nor the non-conventionality of simultaneity. Although I still believe that the relation of distant simultaneity in a single frame of reference is conventional, I am less sure than I once was of the impossibility of rendering it non-conventional without at the same time undermining its relativity to a given reference frame. Perhaps someone will provide an objective method for ascertaining the one-way speed of light.

Although I am convinced that the thesis of conventionality of simultaneity is correct, my main purpose has not been to argue for that view. Rather, I have tried to exhibit the importance of the problem, and to investigate the extent of its ramifications. I will be satisfied if I have made a convincing case for the indispensability of considering the question if we hope for a thorough understanding of the logical foundations of the special theory of relativity. Almost everyone would agree that all physical theories have elements of factual content and elements of conventional decision. The basic philosophical task is, I believe, to try to discover which aspects of a physical theory describe the physical world, and which merely reflect the notations we have chosen for the purposes of that description. A further task, which is also of primary philosophical importance, is to try to ascertain the grounds and status of the various conventions we discover. Some conventions turn out to be strikingly non-trivial; the conventional choice of a value for ϵ (if it is, indeed, a convention) is perhaps the most outstanding example.[48]

SUGGESTED READINGS

1. d'Abro, A. *The Evolution of Scientific Thought*. New York: Dover Publications, Inc., 1950.
2. American Association of Physics Teachers. *Special Relativity Theory*. New York: American Institute of Physics, n.d.
3. Born, Max. *Einstein's Theory of Relativity*. New York: Dover Publications, Inc., 1962.
4. Bridgman, P.W. *A Sophisticate's Primer of Relativity*. Middletown, Conn.: Wesleyan University Press, 1962.
5. Einstein, Albert. *Relativity, The Special and the General Theory*. New York: Henry Holt and Co., 1921. Paperback ed., Crown Publishers, 1968.
6. Einstein, Albert, et. al. *The Principle of Relativity*. New York: Dover Publications, Inc., n.d.
7. Reichenbach, Hans. *The Philosophy of Space and Time*. New York: Dover Publications, Inc., 1957.
8. Taylor, Edwin F., and Wheeler, John Archibald. *Spacetime Physics*. San Francisco: W. H. Freeman and Company, 1966.
9. van Fraassen, Bas C. *An Introduction to the Philosophy of Time and Space*. New York: Random House, 1970.

EPILOGUE

In our attempt to elucidate some of the basic philosophical problems relating to space, time, and motion, we took the work of Thales, the first philosopher and the first real geometer, as our point of departure. He certainly was not the first human to try to solve problems concerning the structure of space, but he seems to have been the first to approach them in a genuinely mathematical way. He got it all started, we might say, not on account of the theorems he proved, but rather because of his realization that geometrical propositions are things that are amenable to proof. Zeno, by contrast, almost brought the whole business to a crunching halt with ingenious arguments designed to prove the impossibility of geometry, as well as the unreality of space, time, and motion. In any case, the ancient Greeks presented us with two gifts. From Zeno we have received problems which have challenged the best philosophical and mathematical minds for 2,500 years—recall from chapter 2 that a new Zeno-type problem has come to light in the 1970s. From Thales, Pythagoras, Euclid, and a host of other contributors we have received powerful tools for dealing with these and other problems. In this connection, it should perhaps be mentioned that Archimedes, in the third century B.C., took us to the very threshold of the infinitesimal calculus. These tools were not, however, completely adequate to meet Zeno's challenge. As we have seen, not only are the analytic geometry of Descartes and the calculus of Newton and Leibniz required, but also subsequent crucial work on the logical foundations of the calculus and modern measure theory.

Although the Greeks possessed remarkable mathematical sophistication—which, to be sure, they applied to certain physical problems — no physics of comparable sophistication existed before the work of Isaac Newton in the seventeenth century. Space, time, and motion were among his most fundamental concerns. As is well known, he expounded doctrines of absolute space, absolute time, and absolute motion. G. W. Leibniz, who shares with Newton the credit for discovering the infinitesimal calculus, attempted to carry through a consistent relativistic conception of space, time, and motion, but I think it must be conceded that Leibniz had no adequate way within his relativistic framework to handle the simple problem of rotational motion. Immanuel Kant, the eighteenth-century philosopher who provided the metaphysical and epistemological foundations of Newtonian science, regarded propositions expressing the structure of space and time as synthetic a priori truths.

We have seen how, in the nineteenth century, the privileged status of Euclidean geometry was challenged, first logically and then epistemologically, by the advent of the non-Euclidean geometries. We have also seen how the absolutistic conception of time was undermined by Einstein's penetrating analysis of the concept of simultaneity — an analysis that lies at the very basis of his special theory of relativity. And we have seen how special relativity forces radical revision of our common-sense notions of space, time, and motion — the common-sense concepts which are largely the result of the long-standing supremacy of Newtonian ideas. Here our story has concluded.

Let us, nevertheless, take an extremely quick glimpse beyond. Einstein's special theory of relativity, first published in 1905, was a steppingstone to his general theory of relativity which appeared in 1916. It is in this context that the non-Euclidean geometries, developed in the nineteenth century, first came to have serious physical application in the description of physical space. We cannot go into the details of general relativity without introducing mathematical techniques that go far beyond the high school algebra and geometry to which we have confined ourselves in this book. But we can take note of the fact that within a short time the general theory of relativity enjoyed considerable success in dealing with a very small number of phenomena to which classical physics and special relativity were quite inadequate. One of these phenomena was the mysterious precession of the perihelion of Mercury, another was the bending of starlight passing close to the sun during solar eclipse, and a third was the so-called "gravitational red-shift."

Until quite recently, however, the general theory had such limited applicability in the explanation and prediction of actually observable phenomena that it was more an object of aesthetic admiration than of practical interest. Moreover, its confirmation was so limited that serious questions about its general correctness were raised. In the last few years, however, chiefly as a result of startling new discoveries in observational astronomy, there has been a tremendously renewed interest in general relativity and cosmology. Many new kinds of empirical evidence have become available, and important new theoretical work has been done.[1] The single most significant occurrence was the discovery by Penzias and Wilson in 1965 of the 3°K cosmic background radiation, which, in the opinion of many experts, has transformed cosmology from speculation into science. Considerable progress toward the empirical ascertainment of the overall space-time structure of the universe has been made. Such phenomena as gravitational collapse, black holes, and gravitational waves lie at the forefront of exciting new developments in astronomy, physics, mathematics, and philosophy. Our story has been an attempt to cover some important aspects of the conceptual groundwork on which these ideas have been erected. The reader who wishes to learn more about recent scientific developments should find the following books both stimulating and helpful.

SUGGESTED READINGS

The following two books, which do not presuppose any extensive scientific background, contain excellent introductions to contemporary work in cosmology and general relativity:

Davies, P. C. W. *Space and Time in the Modern Universe*. Cambridge: Cambridge University Press, 1977.

Weinberg, Steven. *The First Three Minutes*. New York: Basic Books, 1977.

Further references to the philosophical literature on space and time are given in the annotated *BIBLIOGRAPHY* which begins on page 141.

NOTES

Chapter One

1. Euclid's formulation of geometry was not, of course, faultless; indeed, even his deduction of his first proposition is fallacious. The difficulties in Euclid's *Elements* are detailed in the thoroughly annotated Heath edition. Modern axiomatizations of Euclidean geometry that are not vitiated by such errors are given in Hilbert's *The Foundations of Geometry* and in Veblen's article in *Monographs on Topics of Modern Mathematics*, edited by J. W. A. Young. Full bibliographical data on these works will be found in the list of suggested readings at the end of Chapter 1. It is not surprising that Euclid made mistakes; what is absolutely astonishing is the degree of perfection he did attain at a time so early in the history of geometry.
2. Euclid lists five postulates and five axioms or "common notions." These latter assumptions are supposed to be general logical principles—for example, "the whole is equal to the sum of its parts," or "when equals are added to equals the sums are equal"—which hold good in all areas of investigation. The postulates are the assumptions peculiar to geometry. At the present time no significant distinction is made between axioms and postulates; indeed the two terms are virtually synonyms.
3. Actually, Euclid employed a different postulate, but this well-known equivalent, known as "Playfair's Axiom," is more perspicuous in this context.
4. His book was entitled *Euclides ab omni naevo vindicatus* (Milan: P.A. Montani, 1733).
5. The geometry of Bolyai and Lobachevski is actually a geometry of constant curvature, as are Euclidean geometry and Riemann's geometry of the sphere. The pseudo-sphere has a surface of constant curvature, but the saddle does not. Nevertheless, the saddle surface is useful for illustrative purposes because of its familiarity, and its geometry does approximate that of the pseudo-sphere. Riemann, incidentally, in addition to discovering the geometry of constant positive curvature, also developed differential geometry—a highly generalized treatment of a wide variety of geometries using techniques of the infinitesimal calculus. Differential geometry includes geometries in which the curvature varies from place to place. The world with the hump (see Figure 7) involves a geometry of variable curvature. For an excellent introduction to differential geometry see A.D. Aleksandrov, A.N. Kolmogorov, M.A. Lavrent'ev, eds., *Mathematics: Its Content, Methods, and Meaning* (Cambridge, Mass.: The M.I.T. Press, 1963), Chap. XVII, § 9.
6. The concept of curvature is a well-defined mathematical concept; see Adler, *A New Look at Geometry*, for an introductory account.
7. Henri Poincaré's classic *Science and Hypothesis* presents a very readable account.

8. The denial of a self-contradiction is an analytic statement, and analytic statements are vacuous—they have no factual content. Since Kant maintained that the propositions of geometry are informative, it was essential for him to avoid treating them as analytic.
9. The entire Lobachevskian two-dimensional space cannot be embedded in three-dimensional Euclidean space. The surface of the pseudo-sphere as pictured in Figure 3 is merely part of such a space. But, it is well to recall, the surface of a chalkboard represents only part of the infinite Euclidean plane.
10. Even if the space of our universe is non-Euclidean, it so closely approximates a Euclidean space in terrestrial dimensions that our everyday experience is indistinguishable from what it would be in a strictly Euclidean world.
11. The distinction between internal and external visualization parallels a distinction between what might be called internal and external curvature. Gaussian curvature, which is the type of curvature that characterizes the geometry of the sphere, the plane, and the pseudo-sphere can be rigorously defined in terms of internal properties of the space. Gaussian curvature need *not* be defined in relation to a space of higher dimension in which the surface is embedded. Other types of curvature, for example, mean curvature, are defined in terms of lack of conformity to a plane in a three-dimensional Euclidean embedding space. The surface of a right circular cylinder has mean curvature, but its Gaussian curvature is zero. Its geometry is Euclidean as you can see by drawing geometrical figures (such as circles and triangles) on a flat piece of paper. When this is rolled up into a cylinder, none of the metrical relations within the figures are changed.
12. As Lobachevski himself realized, measurements of stellar parallax could reveal non-Euclidean properties of space. In effect, the astronomer measures two angles of a triangle and subtracts their sum from 180°. If the result were always positive even for stars indefinitely far away, that would be evidence for a Lobachevskian space. If the result were sometimes negative, that might indicate a Riemannian space. However, this latter result must be treated with caution, for negative results can arise from certain observational errors.
13. See Hans Reichenbach, *The Philosophy of Space and Time* (New York: Dover Publications, 1958) § 9-13, for a more detailed discussion of visualization.
14. The distinction between differential and universal forces is due to Reichenbach. This terminology has led to the criticism that such forces are introduced ad hoc. However, the term "force" in this context is metaphorical. Literally speaking, there are no such "forces" at all, there are only different definitions of congruence. Rudolf Carnap, *Philosophical Foundations of Physics* (New York: Basic Books, 1966) suggests that "differential effects" and "universal effects" are less misleading terms. The term "effect" is equally metaphorical in this context.
15. Adolf Grünbaum has investigated the concept of an intrinsic metric and its relation to the geometry of space in considerable detail. See his "Space, Time and Falsifiability" *Philosophy of Science*, XXXVII, 4 (Dec. 1970), pp. 469-588. Reprinted in Grünbaum, *Philosophic Problems of Space and Time*, 2nd ed., chap. 16.

NOTES 133

16. If space had a discrete structure it might well possess an intrinsic metric, but we are not considering this possibility here because the types of geometry we are discussing all assume the continuity of lines, surfaces, and so on. In the next chapter we shall raise some issues related to discrete spaces.
17. The theory of equivalent descriptions was elaborated at length by Reichenbach in a number of his works.
18. In former times the meter was defined as the distance between two marks on the standard meter, a platinum-iridium bar located near Paris. Other meter sticks could be standardized by comparison with the standard meter. But the question which could only be answered by a further stipulation was whether these meter sticks retained their length when transported away from Paris. Nowadays, the meter is defined in terms of a certain number of wavelengths of light in a certain spectral line, but essentially the same question remains. Does that wavelength remain constant regardless of its location?

Chapter Two

1. For further historical information and references see, Wesley C. Salmon, *Zeno's Paradoxes* (Indianapolis: The Bobbs-Merrill Company, Inc., 1970). The "Introduction" contains a general discussion of these paradoxes and their implications. For extended systematic discussion of these paradoxes see Adolf Grünbaum, *Modern Science and Zeno's Paradoxes* (Middletown, Conn.: Wesleyan University Press, 1967).
2. Bertrand Russell, *Our Knowledge of the External World* (New York: W. W. Norton & Company, Inc., 1929), p. 193. The main portion of the discussion of Zeno is reprinted in Salmon, *Zeno's Paradoxes*.
3. These formulations are taken from Salmon, *op. cit.*, pp. 8-12.
4. Russell, *op. cit.*, p. 189.
5. See J. O. Wisdom, "Achilles on a Physical Racecourse," reprinted in Salmon, *Zeno's Paradoxes*.
6. See G. E. L. Owen, "Zeno and the Mathematicians" reprinted in Salmon, *Zeno's Paradoxes*.
7. Charles Hartshorne and Paul Weiss, eds. *The Collected papers of Charles Sanders Peirce* (Cambridge, Mass.: Harvard University Press, 1935) § 6.177-184.
8. For an excellent discussion of these developments see Carl B. Boyer, *The History of the Calculus and its Conceptual Development* (New York: Dover Publications, Inc., 1959).
9. Zeno's paradoxes pose enormous problems in historical scholarship; for some of the details see Gregory Vlastos, "Zeno of Elea," in The *Encyclopedia of Philosophy*, ed. Paul Edwards (New York: The Macmillan Company and The Free Press, 1967).
10. Reprinted in James R. Newman, ed., *The World of Mathematics* (New York: Simon and Schuster, 1956).
11. Henri Bergson, *Creative Evolution*, trans. Arthur Mitchell (New York: Holt,

Rinehart and Winston, 1911), relevant passages reprinted in Salmon, *Zeno's Paradoxes*, quotation p. 63.

12. Bertrand Russell, *The Principles of Mathematics*, 2nd ed. (New York: W. W. Norton & Company, Inc., 1943), p. 347.
13. The contrary view, that this is indeed an absurdity, is based upon the elementary fallacy of composition. This is the only non-trivial, non-artifical instance of this fallacy I have ever encountered.
14. For detailed and enlightening discussions of the relations between "physical time" and "psychological time," see Adolf Grünbaum, "Relativity and the Atomicity of Becoming," *Review of Metaphysics*, IV (1950-51), 143-186.
15. Max Black, "Achilles and the Tortoise," *Analysis* XI (1950-51), 91-101; reprinted in Salmon, *Zeno's Paradoxes*.
16. *Ibid.*, p. 72 in Salmon.
17. The idea of an infinity machine was first suggested by Hermann Weyl, *Philosophy of Mathematics and Natural Science* (Princeton, N. J.: Princeton University Press, 1949). See Salmon, *Zeno's Paradoxes*, p. 201, for relevant quotation.
18. James Thomson, "Tasks and Super-Tasks," *Analysis* XV (1954-55), 1-13; reprinted in Salmon, *Zeno's Paradoxes*.
19. The "switching function" may be defined as follows: let 1 represent the "on-state" of the lamp, and let 0 represent the "off-state." This function has a determinate value for each value of $t < 1$, but it fluctuates infinitely often between 0 and 1 in any neighborhood of $t = 1$; hence, it has no limit at $t = 1$.
20. The arguments of this paragraph were given by Paul Benacerraf, "Tasks, Super-Tasks, and the Modern Eleatics," *Journal of Philosophy* LIX (1962), 765-84; reprinted in Salmon, *Zeno's Paradoxes*.
21. This analysis of infinity machines and their modifications is due to Adolf Grünbaum, "Modern Science and Zeno's Paradoxes of Motion," Part II, in Salmon, *Zeno's Paradoxes*, pp. 218-44.
22. *Ibid.*, Part I, "The Zenonian Runners", pp. 204-18.
23. See Salmon, *Zeno's Paradoxes*, pp. 215-16, for the details.
24. *Scientific American*, July, 1971.
25. A "supertask" is simply an infinite set of ordinary tasks.
26. *Scientific American*, December, 1971, pp. 97-99.
27. See Salmon, *Zeno's Paradoxes*, pp. 12-15, for a full statement of this paradox.
28. See H.D.P. Lee, *Zeno of Elea* (Cambridge: University Press, 1936), pp. 12-13 (fragment 2); p. 22.
29. This analysis of the paradox of plurality is due to Adolf Grünbaum, "Zeno's Metrical Paradox of Extension" in Salmon, *Zeno's Paradoxes*, pp. 176-99.
30. *Ibid.* For brief and clear expositions of the fundamentals of measure theory see J.C. Burkill, *The Lebesgue Integral* (Cambridge: University Press, 1953) or Harald Cramér, *Mathematical Methods of Statistics* (Princeton University Press, 1946), Chaps. 1-7.

31. The series $1 - 1/2 + 1/3 - 1/4 + 1/5$ ... is not absolutely convergent, that is, the series $1 + 1/2 + 1/3 + 1/4$ ... consisting of the absolute values of the terms of the foregoing series does not converge. Series which converge, but do not converge absolutely, may suffer a change in sum, or even become divergent, if their terms are rearranged.
32. For a detailed discussion of metrical amorphousness and intrinsic metrics, see Adolf Grünbaum, "Space, Time and Falsifiability," *Philosophy of Science*, XXXVII (Dec., 1970), pp. 469-588. Reprinted in Grünbaum, *Philosophic Problems of Space and Time*, 2nd ed., chap. 16.
33. In some respects the one-dimensional case is a serious over-simplification of the two- or higher-dimensional case. In the one-dimensional case there can be no such thing as internal curvature, so the geometry of any line is Euclidean. This means that it is always feasible to equate distance with coordinate differences. In the higher-dimensional cases we need metric coefficients (analogous to what we are calling metric rules), and these determine entirely the *internal curvature*, and hence the geometry, of the space. For details see A. D. Aleksandrov, "Non-Euclidean Geometry" in A. D. Aleksandrov, A. N. Kolmogorov, and M. A. Lavrent'ev, *Mathematics: Its Content, Methods, and Meaning* (Cambridge, Mass.: The M.I.T. Press, 1963, 1963), § 9, "Riemannian Geometry."
34. A continuous function is, intuitively, one that can be plotted by means of a line that has no gaps in it—one that can be drawn without lifting the pencil from the paper on which the function is being plotted. For a respectable mathematical treatment of the concept of continuity, in terms requiring no previous mathematical training beyond high school, see Richard Courant and Herbert Robbins, *What is Mathematics?* (New York: Oxford University Press, 1941), Chap. VI.
35. P. W. Bridgman, "Some Implications of Recent Points of View in Physics," *Revue Internationale de Philosophie*, III (1949), p. 490; quoted by Grünbaum, see Salmon, *Zeno's Paradoxes*, p. 177.
36. See Salmon, *Zeno's Paradoxes*, pp. 16-20, for discussion of metaphysical interpretations of these paradoxes; see pp. 59-66 for a famous passage from Bergson.
37. See Grünbaum, "Modern Science and Zeno's Paradoxes of Motion," Part III, in Salmon, *Zeno's Paradoxes*, pp. 244-60, for an assessment of the extent to which the quantization of space and time has been accomplished.
38. Weyl, *op. cit.*, p. 43. See Salmon, *Zeno's Paradoxes*, p. 175, for the relevant quotation.
39. "Weyl" is pronounced like "vile."
40. See Peter D. Asquith, *Alternative Mathematics and Their Status*. Ph.D. dissertation, Indiana University, 1970.

Chapter Three

1. In this chapter I shall attempt to present a few basic features of Einstein's special theory of relativity that are needed to understand the following chapter. More detailed presentations, still at an elementary level, are available in many sources. *Einstein's Theory of Relativity* by Max Born (New

York: Dover Publications, Inc., 1962) and *Relativity, The Special and General Theories* by Albert Einstein (New York: Henry Holt and Co., 1921; paperback, Crown, 1968) are classics; *Spacetime Physics* by Edwin F. Taylor and John Archibald Wheeler (San Francisco: W. H. Freeman and Company, 1966) is a lively modern presentation with many interesting examples and illustrations. Most contemporary introductory physics texts contain at least a chapter on special relativity. This chapter differs from many treatments by emphasizing the relationship between the relativity of simultaneity on the one hand and the Lorentz contraction and time dilation on the other.

2. An account of the Michelson-Morley experiment can be found in almost any introductory treatment of special relativity, including those mentioned in the preceding note. A good discussion is given in *Introduction to Concepts and Theories in Physical Science* by Gerald Holton, 2nd edition revised and supplemented by Stephen G. Brush (Reading, Mass.: Addison-Wesley Publishing Co., 1973), chap. 30.
3. Albert Einstein, "On the Electrodynamics of Moving Bodies," in Einstein, et al., *The Principle of Relativity* (New York: Dover Publications, Inc., n.d.).
4. An inertial reference frame is one in which bodies not subject to external forces move in straight lines at uniform velocity—in short, a reference frame in which Newton's first law of motion holds.
5. Einstein's "Autobiographical Notes" in P.A. Schilpp, ed., *Albert Einstein: Philosopher-Scientist* (Evanston, Ill.: The Library of Living Philosophers, Inc., 1949), p. 53.
6. Albert Einstein, *Relativity, The Special and The General Theory.*
7. The fact that distant simultaneity requires some such definition—as well as the grounds for choosing this particular definition—constitute the central issue of the next chapter.
8. Given that we have adopted the definition of simultaneity given above. The conventionality inherent in that definition will be discussed in the next chapter. It is perhaps worth emphasizing that there is nothing bizarre about the definition we have adopted; it is a standard part of every treatment of the special theory of relativity.
9. We could obviously have chosen our ground observer as the origin of our coordinate system, in which case the t-axis would have been his world line. The rock where lightning bolt 1 struck was chosen instead in order to lay a foundation for figures 4 and 5.
10. In this fanciful situation, where trains travel at 6/10ths of the speed of light, we shall not boggle at birds which fly at 4/10ths of the speed of light.
11. People sometimes become confused about units of measure of time or distance based upon the velocity of light. The "light year," for example, is a measure of spatial distance—the *distance* light travels in a year's time. Our "light-meter" is a measure of time duration—the amount of *time* required for light to travel one meter. Since a velocity is a distance divided by a time

$$v = d/t$$

NOTES 137

multiplying a velocity by a time yields a distance (light year),

$d = vt$

while dividing a distance by a velocity yields a time (light-meter)

$t = d/v$

12. The photon is a quantum mechanical entity, but our discussion will make no reference to any of its quantum mechanical properties. For our purposes it is simply the smallest pulse of light available.
13. As we shall see in the next chapter, these assumptions about the velocity of the train relative to the ground and the velocity of the ground relative to the train are by no means trivial; for now, we shall merely take them as given.
14. The length contraction occurs only in the dimension parallel to the direction of motion. Lengths oriented perpendicular to the direction of motion—for example, the length of the train observer's light clock when it is held in a vertical position—are unaffected. They are the same in both the stationary and the moving reference frames.
15. See Edwin F. Taylor and John Archibald Wheeler, *Spacetime Physics* (San Francisco: W. H. Freeman & Co., 1963), pp. 16-17.

Chapter Four

1. Then of Washington University, but more recently—ironically—with the Caterpillar Tractor Company of Peoria, Illinois.
2. I would prefer to call this the "twin paradox," reserving the term "clock paradox" for the problem to be discussed below under that name.
3. J. S. Hafele and Richard E. Keating, "Around-the-World Atomic Clocks: Predicted Relativistic Time Gains and Observed Relativistic Time Gains," *Science*, 177, pp. 166-70 (14 July 1972).
4. See Edwin F. Taylor and John Archibald Wheeler, *Spacetime Physics* (San Francisco: W. H. Freeman and Company, 1963), p. 89.
5. The cesium beam clock is an "atomic clock" but it is nevertheless a macroscopic object. The fact that it has atomic parts does not distinguish it from other macroscopic objects, including human beings.
6. W. P. Montague, *Philosophical Review* XXIII (1924), pp. 156ff; A. O. Lovejoy, *Philosophical Review* XL (1931) pp. 48, 152, 549.
7. Of course, both of them might experience accelerations, but this case only introduces further complications without adding anything of interest to the problem.
8. At time of writing: Yesterday, Art Pollard was killed at the Indianapolis Speedway when his car hit a concrete wall.
9. Historical note: John Cameron Swayze is a TV announcer who narrated many commercials in which Timex watches were subjected to severe accelerations (for example, being attached to the ski tip of an Olympic ski

jumper during a jump) without impairment of function. At the conclusion of each such demonstration Swayze would triumphantly proclaim of Timex watches that they can "take a licking and keep on ticking." Some years earlier, when home permanents were being introduced to the public via television, the manufacturers of Toni home permanents had a series of commercials in which one twin had a permanent by a professional hairdresser while the other had a Toni home permanent. The motto, "Which twin has the Toni?" was designed to suggest indistinguishability.

10. Adolf Grünbaum, "The Clock Paradox in the Special Theory of Relativity" *Philosophy of Science* XII (1954), pp. 249-53.

11. By H. Bondi, "The Space Traveller's Youth," *Discovery*, Dec. 1957, pp. 505-10. This formulation is also employed by Grünbaum, *op. cit.*

12. These clock readings can be computed as follows: In the interval t''_0 to t''_3 clock C'' registers an elapsed time of $(2d/v)\sqrt{1-\beta^2}$. Compared with C the time of C'' is dilated by a factor $\sqrt{1-\beta^2}$. Hence

$$t_3 - t^* = \frac{2d}{v}(1-\beta^2)$$

$$t_3 = \frac{2d}{v}$$

$$t^* = \frac{2d}{v}\beta^2$$

Furthermore, in the interval t''_1 to t''_3 C registers an elapsed time of $2d/v$. Since the time of C is dilated by the factor $\sqrt{1-\beta^2}$ with respect to K'' the interval

$$t''_3 - t''_1 = \frac{2d}{v} \cdot \frac{1}{\sqrt{1-\beta^2}}$$

$$t''_3 = \frac{2d}{v}\sqrt{1-\beta^2}$$

$$t''_1 = \frac{2d}{v}\sqrt{1-\beta^2} - \frac{2d}{v} \cdot \frac{1}{\sqrt{1-\beta^2}}$$

$$= \frac{2d(1-\beta^2)}{v\sqrt{1-\beta^2}} - \frac{2d}{v\sqrt{1-\beta^2}}$$

$$= -\frac{2d}{v}\beta^2 \frac{1}{\sqrt{1-\beta^2}}$$

13. This calculation is carried out in Grünbaum, *op. cit.*, but the crucial calculation, from the frame of C'', is surprisingly omitted. Hence, the analysis just presented is merely the completion of the work done earlier by Grünbaum.

14. In P. A. Schilpp, ed., *Albert Einstein: Philosopher Scientist* (Evanston, Ill.: The Library of Living Philosophers, Inc., 1949).

15. "On the Electrodynamics of Moving Bodies," in Einstein, *et al.*, *The Principle of Relativity* (New York: Dover Publications, n.d.), p. 40.

NOTES

16. *Ibid.*
17. Certain pseudo-causal processes can travel faster than light, but they are useless for transmission of information. See Hans Reichenbach, *The Philosophy of Space and Time* (New York: Dover Publications, Inc., 1957) § 23 "Unreal Sequences."
18. Michael N. Kreisler, "Are There Faster-than-Light Particles?" *American Scientist* LXI (1973), pp. 201-208. We shall return to tachyons below, p. 122.
19. Reichenbach, *Philosophy of Space and Time*, pp. 126f. Reichenbach talks about simultaneity rather than synchrony in this passage, but this is an inconsequential matter of terminology, for clocks are synchronized if they show the same readings simultaneously.
20. *Ibid.*, p. 127.
21. Hans Reichenbach, "Planetenuhr und Einsteinsche Gleichzeitigkeit" *Zeitschrift für Physik* XXXIII (1924), pp. 628-34, provides a detailed discussion of Römer's measurement.
22. P. W. Bridgman. *A Sophisticate's Primer of Relativity* (Middletown, Conn.: Wesleyan University Press, 1962), pp. 64-67.
23. Brian Ellis and Peter Bowman, "Conventionality in Distant Simultaneity," *Philosophy of Science*, 34, 2(June, 1967), pp. 116-36.
24. Adolf Grünbaum, Wesley C. Salmon, Bas C. van Fraassen, and Allen I. Janis, "A Panel Discussion of Slow Clock Transport in the Special and General Theories of Relativity," *Philosophy of Science* 36, 1 (March, 1969) pp. 1-81. The contribution by Janis deals with general relativity.
25. "The Conventionality of Simultaneity," *ibid.*
26. *Ibid.*, p. 61.
27. Reichenbach, *Philosophy of Space and Time* § 20, "Attempts to determine Absolute Simultaneity." The original German edition, *Philosophie der Raum-Zeit-Lehre*, was published in 1927.
28. One such proposal was made by H. Weinberger and M. Mossel "Theory of a Unidirectional Interferometric Test of Special Relativity," *American Journal of Physics*, 39 (June, 1971) but it was shown invalid by G.E. Stedman, "A Unidirectional Test of Special Relativity?" *American Journal of Physics*, 40 (May, 1972).
29. See Reichenbach, *The Philosophy of Space and Time*, p. 147f, for discussion of this example.
30. R.S. Shankland, "Conversations with Albert Einstein," *American Journal of Physics*, XXXI (1963).
31. This is shown by Ellis and Bowman *op. cit.*, pp. 123f, as well as by John Winnie in the paper I am about to discuss.
32. See Adolf Grünbaum, *Philosophical Problems of Space and Time* (New York: Alfred A. Knopf, 1963; 2nd ed., Boston: D. Reidel Publishing Co., 1974), pp. 359-68.
33. "Special Relativity Without One-Way Velocity Assumptions," *Philosophy of Science*, 37, 1-2 (March and June, 1970), pp. 81-99; 223-238.
34. *Ibid.*, pp. 229-31.

35. This principle, like the reciprocity condition, is sufficient for the derivation of $\epsilon = 1/2$. See Winnie, *ibid.*, p. 230.
36. *Ibid.*, p. 231n.
37. *Ibid.*, pp. 231-37.
38. *Ibid.*, pp. 91-93.
39. *Ibid.*, pp. 97-97.
40. *Ibid.*, pp. 97-98. See also John Winnie, "The Twin-Rod Thought Experiment," *American Journal of Physics*, 40 (August, 1972), pp. 1091-94.
41. Winnie, "Special Relativity Without One-Way Velocity Assumptions," pp. 223-28.
42. See Max Born, "Physical Reality," *Philosophical Quarterly*, 1953, pp. 139-49, for an excellent discussion of the philosophical significance of invariants in physical theories. Reprinted in Max Born, *Physics in My Generation*, 1st ed., Pergamon Press.
43. Gerald Feinberg, *Physical Review*, 159 (1967) p. 1089.
44. This example is due to Roger G. Newton, "Particles That Travel Faster than Light?" *Science*, vol. 167, pp. 1569-1574 (20 March, 1970).
45. These concepts are explained in a clear and elementary manner in Taylor & Wheeler, *Spacetime Physics*.
46. For a clear and charming exposition of this procedure, see "The Grandfather Clock in Outer Space" in Leon Cooper, *An Introduction to the Meaning and Structure of Physics* (New York: Harper and Row, 1968), pp. 401-406. Cooper spells out in some detail the inconveniences and disadvantages of this approach to simultaneity.
47. See Salmon, "The Conventionality of Simultaneity," § 1.
48. See Salmon, "The Conventionality of Simultaneity," for discussion of the nature of this nontriviality.

Epilogue

1. A striking recent example is the discovery of a gravitational lens, a phenomenon whose possibility is a direct consequence of general relativity, but which had not been detected before 1979. For an extremely readable account of this fascinating example, see Frederic H. Chaffee, Jr., "The Discovery of a Gravitational Lens," *Scientific American*, 243 (November, 1980), pp. 70–88.

BIBLIOGRAPHY

The following annotated list contains selected writings in the philosophy of space and time. To keep the size within reasonable limits, I have not included any references to the large and excellent *scientific* literature in mathematics, physics, and cosmology that has direct bearing upon the *philosophical* issues raised in this book. Some works in the latter category that seem especially relevant, or likely to be helpful, have been mentioned in the *suggested readings* at the ends of the chapters.

In this book, I have endeavored to provide a basic introduction, which does not presuppose any technical background beyond high school mathematics, to problems in the philosophy of space and time. The standpoint I have adopted is the classic approach, anticipated by Henri Poincaré at the beginning of the present century, developed and advocated primarily by Hans Reichenbach in the first half of this century, and continued and extended chiefly by Adolf Grünbaum since 1950. Needless to say, not all philosophers find this viewpoint acceptable. In the large—and sometimes rather technical—literature on the philosophy of space and time, there is considerable controversy. It is my opinion that, whatever view one finally adopts, an understanding of the Reichenbach-Grünbaum approach is a necessary prerequisite to an appreciation of the problems. My main aim in writing this book has been to help to provide an *entre* into this area, which constitutes an extremely significant field of current philosophy of science, as well as a fundamental component in the development of the entire field of philosophy from the time of the ancient Greeks right up to the present day.

The reader who wishes to pursue the philosophical issues raised in this book—by studying them in more detail, by investigating other closely related topics, or by examining alternative viewpoints—will find many valuable works listed in the bibliography below. Before giving the formal bibliography, however, I shall offer some comments on leading works that fall into several important categories. I believe these remarks will be a helpful supplement to the bibliography.

ELEMENTARY INTRODUCTIONS

While I do not know of any other book that approaches these philosophical topics on as elementary a level as this one, parts II-III of Carnap (1966) and chapters 8-9 of Reichenbach (1951) contain very elementary introductions by two of the most important contributors to this classic approach. Reichenbach's (1949) essay is also quite elementary. Davies (1977) surveys pertinent issues from a more scientific—though still elementary—point of view.

Smart (1964) and Gale (1967) are good introductory anthologies.

Introductions that are somewhat more extended and more technical can be found in Sklar (1974) and van Fraassen (1970). Van Fraassen contains a great deal of interesting historical material.

MAJOR CLASSICS

An early classic (first published in English in 1905) that has stimulated much discussion is Poincaré (1952). The major classic of the first half of the 20th century is Reichenbach's systematic treatise (1956), first published in German (1928). Also important is Reichenbach's early attack upon the Kantian viewpoint (1965), first published in German (1920), as well as his axiomatization of relativity theory (1969), first published in German (1924).

Torretti (1978) provides a detailed discussion of historical developments in the latter half of the 19th century. Čapek (1976) is an anthology containing a great deal of historical material.

At present, Grünbaum (1973) stands as the unrivaled systematic treatise on the philosophy of space and time in the second half of the 20th century. A large portion of the material in Earman, et al. (1977) is in response to Grünbaum's work.

COLLECTIONS OF RECENT RESEARCH

Suppes (1973), Earman, et al. (1977), and Winnie (1977a) are collections of significant recent essays in the philosophy of space and time; many are fairly technical. Widely divergent points of view are expressed by the various authors who have contributed to these volumes.

SOME IMPORTANT CONTROVERSIES

Putnam (1963) offers a spirited attack on Grünbaum's views on *space and geometry*; Grünbaum (1968), chapter III, contains a vigorous rebuttal. Friedman (1973) and Fine (1973) contain further critical discussions of these issues.

Nagel (1961), chapters 8-9, and Ellis (1963) contain criticisms of Reichenbach's views on *universal forces* and the *definition of congruence*.

Ellis and Bowman (1967) contains an attack on the Reichenbach-Grünbaum thesis of the *conventionality of simultaneity*. A detailed answer is given by Grünbaum, et al. (1969). Winnie (1970) provides an importantly different response. Ellis (1971) and Bowman (1977) contain rejoinders to Grünbaum, et al. Salmon (1977) contains further extended discussion of this issue. Malament (1977) furnishes an original contribution of the first importance regarding the conventionality of simultaneity. Further discussion of these issues—opposing the Reichenbach-Grünbaum viewpoint—will be found in Nerlich (forthcoming) and in Saunders and Norton (forthcoming).

Graves (1971) is a philosophical apology for *geometrodynamics*, a view expounded by Wheeler (1962)—indeed, Wheeler contributed a Foreword to Graves. This geometrodynamic approach is in fundamental disagreement with the conclusions of Reichenbach and Grünbaum, as Grünbaum explains in detail in (1973), chapter 22, and in (1973a), where he takes basic issue with geometrodynamics.

The fundamental issue separating Grünbaum from Wheeler concerns the question of whether space has *intrinsic curvature* (as Wheeler requires) or whether it is *metrically amorphous* (as Grünbaum maintains). This issue is discussed in detail in Grünbaum (1973), chapter 16, where he considers many criticisms of his doctrine of metrical amorphousness. In Salmon (1977a), I attempt to answer a technical criticism by Glymour (1972). Nerlich (1979) is a rebuttal to my (1977a). Nerlich (1976) is a systematic exposition of the view that physical space has an intrinsic geometric structure.

ALTERNATIVE APPROACHES

There are several systematic treatises that adopt viewpoints which differ radically from those of Reichenbach and Grünbaum, and from that which I have presented in this book. As mentioned above, both Graves (1971) and Nerlich (1976) defend the view that physical space has intrinsic geometrical characteristics. Swinburne (1968) expounds an ordinary language approach to space and time, as does Gale (1968) to time. Čapek (1961) adopts an approach that is close to the metaphysical doctrines of Bergson and Whitehead. Whiteman (1967) approaches space and time from a standpoint which is, roughly speaking, phenomenological. Russell (1897) attempted to defend a Kantian viewpoint, even in the face of non-Euclidean geometries.

THE CAUSAL THEORY OF SPACE AND TIME

Although both Reichenbach (1956, §3) and Grünbaum (1973, chap. 8) hold causal theories of time, neither of them would go so far as to maintain that the entire structure of Minkowski spacetime is determined causally. Using results that were first developed by A. A. Robb (1914, 1921, 1936), Winnie (1977) develops a fully causal theory of Minkowski spacetime. It is within just such a framework that Malament (1977) demonstrates the unique causal definability of simultaneity. Sklar (1977) contains extended criticisms of Winnie's approach, and Sklar (1979) contains further criticisms of Reichenbach's causal theory of time.

THE DIRECTION OF TIME

One fascinating issue regarding the nature of time, which was not broached in this book, concerns the basis for the asymmetry (or, as Grünbaum prefers to say, the "anisotropy") of time. Davies (1977, chap. 3) provides a brief introduction to this topic at a nontechnical level. Reichenbach's unfinished posthumous work (1956) is a seminal treatment of this problem, and Grünbaum (1973, chaps. 8, 19) provides further discussion. Davies (1974) is a comprehensive discussion that takes account of scientific results that were not available at the time of Reichenbach's death. Grünbaum (1967, chap. 1) and (1973, chap. 10) shows that, Reichenbach and others to the contrary notwithstanding, temporal anisotropy does not imply that time "flows" from past to future, or that temporal becoming is an objective feature of the physical world.

ANNOTATED LIST OF WORKS:

d'Abro, A., 1950: *The Evolution of Scientific Thought*, 2nd ed. New York: Dover Publications.
    A popular historical discussion of developments leading to
    Einstein's special and general theories of relativity.

Beauregard, Laurent A., 1979: "Reichenbach and Conventionalism," in Wesley C. Salmon, ed., *Hans Reichenbach: Logical Empiricist* (Dordrecht: D. Reidel Publishing Co.), pp. 305–320.
   A discussion of various conventionality claims and their interrelations.

Benardete, José, 1964: *Infinity: An Essay in Metaphysics*. Oxford: Clarendon Press.
   A discussion of many issues directly related to Zeno's paradoxes.

Bowman, Peter, 1977: "On Conventionality and Simultaneity—Another Reply," in Earman, et al. (1977), pp. 433–447.
   An answer to Grünbaum, et al. (1969).

Bridgman, P. W., 1962: *A Sophisticate's Primer of Relativity*. Middletown, Conn.: Wesleyan University Press. With a Prologue and an Epilogue by Adolf Grünbaum.
   A consideration of fundamental philosophical issues by an eminent physicist. Contains a discussion of synchrony by slow clock transport.

Čapek, Milič, 1961: *The Philosophical Impact of Contemporary Physics*. Princeton, N.J.: D. Van Nostrand Co.
   A comprehensive treatise on space and time written from a metaphysical standpoint.

———, 1976: *The Concepts of Space and Time*. Dordrecht: D. Reidel Publishing Co.
   An anthology containing many important historical selections.

Carnap, Rudolf, 1966: *Philosophical Foundations of Physics* (ed., Martin Gardner). New York: Basic Books. Reprinted as *Introduction to the Philosophy of Science*, Harper Torchbook, 1974.
   Parts II–III contain a lucid and readable introduction to issues in the philosophy of space and time.

Coffa, J. Alberto, 1979: "Elective Affinities: Weyl and Reichenbach," in Wesley C. Salmon, ed., *Hans Reichenbach: Logical Empiricist* (Dordrecht: D. Reidel Publishing Co.), pp. 267–304.
   A historical discussion of the status of the affine connection.

Costa de Beauregard, O., 1979: "Two Lectures on the Direction of Time," in Wesley C. Salmon, ed., *Hans Reichenbach: Logical Empiricist* (Dordrecht: D. Reidel Publishing Co.), pp. 341–366.
   Discussions of the problem of temporal anisotropy in classical statistical mechanics and quantum mechanics.

Davies, P. C. W., 1974: *The Physics of Time Asymmetry*. Berkeley: University of California Press.
   A physically sophisticated treatment of the anisotropy of time.

———, 1977: *Space and Time in the Modern Universe*. Cambridge: Cambridge University Press.

A comprehensive elementary discussion of many aspects of
space, time, relativity, and cosmology.

Earman, John, 1977: "How to Talk about the Topology of Time," *Nous*, vol. 11, no. 3, pp. 211–226.
A discussion of the philosophical implications of recent mathematical work on relativity.

Earman, John S., Clark N. Glymour, and John J. Stachel, eds., 1977: *Foundations of Space-Time Theories, Minnesota Studies in the Philosophy of Science*, vol. VIII. Minneapolis: University of Minnesota Press.
An important collection of research papers at the frontiers of philosophy of space and time.

Ellis, Brian, 1963: "Universal and Differential Forces," *British Journal for the Philosophy of Science*, vol. 14, no. 55, pp. 177–194.
An argument against Reichenbach's views on the definition of congruence.

———, 1971: "On Conventionality and Simultaniety—A Reply," *Australasian Journal of Philosophy*, vol. 49, no. 2, pp. 177–203.
A rebuttal to Grünbaum, et al. (1969).

Ellis, Brian, and Peter Bowman, 1967: "Conventionality in Distant Simultaneity," *Philosophy of Science*, vol. 34, no. 2, pp. 116–136.
A provocative challenge to the thesis of conventionality of simultaneity.

Fine, Arthur, 1973: "Reflections on a Relational Theory of Space," in Suppes (1973), pp. 234–267.
A critical discussion of the Putnam-Grünbaum debate.

Friedman, Michael, 1973: "Grünbaum on the Conventionality of Geometry," in Suppes (1973), pp. 217–233.
A vigorous attack on Grünbaum's version of conventionality.

Gale, Richard M., ed., 1967: *The Philosophy of Time*. Garden City, N.Y.: Doubleday & Co.
An anthology providing a broad selection of philosophical writings on time.

———, 1968: *The Language of Time*. London: Routledge & Kegan Paul.
An approach to philosophical problems about time by way of linguistic analysis.

Glymour, Clark, 1972: "Physics by Convention," *Philosophy of Science*, vol. 39, no. 3, pp. 322–340.
A lively criticism of Grünbaum's views on conventionality.

———, 1977: "The Epistemology of Geometry," *Nous*, vol. 11, no. 3, pp. 227–252.
An argument against Reichenbach's theory of equivalent descriptions as it applies to our knowleage of the structure of physical space.

Graves, John Cowperthwaite, 1971: *The Conceptual Foundations of Contemporary Relativity Theory*. Cambridge, Mass.: MIT Press.

A philosophical apology for Wheeler's geometrodynamics, and
criticism of the views of Reichenbach and Grünbaum. Contains a Foreword by John Archibald Wheeler.

Grünbaum, Adolf, 1962: "Geometry, Chronometry, and Empiricism," in Herbert Feigl and Grover Maxwell, eds., *Scientific Explanation, Space, and Time, Minnesota Studies in the Philosophy of Science*, vol. III (Minneapolis: University of Minnesota Press), pp. 405–526.

This monographic essay was the precursor to Grünbaum (1963), and the target of Putnam (1963).

———, 1963: *Philosophical Problems of Space and Time*. New York: Alfred A. Knopf.

A comprehensive treatise on space and time, much expanded in Grünbaum (1973).

———, 1967: *Modern Science and Zeno's Paradoxes*. Middletown, Conn.: Wesleyan University Press. British edition, somewhat revised, London: George Allen and Unwin Ltd., 1968.

The most comprehensive and systematic treatise on Zeno's paradoxes of plurality and motion.

———, 1968: *Geometry and Chronometry in Philosophical Perspective*. Minneapolis: University of Minnesota Press.

Chapter I is a reprint of Grünbaum (1962). Chapter III is a detailed response to Putnam's (1963) attack on Grünbaum (1962). Chapter II is devoted mainly to a discussion of the meaning-status of the claim that everything doubled in size; this discussion was prompted by Schlesinger (1964 and 1967).

———, 1973: *Philosophical Problems of Space and Time*, 2nd enlarged ed. Dordrecht: D. Reidel Publishing Co.

The most influential and comprehensive treatise on philosophy of space and time since Reichenbach's classic works. This second edition of Grünbaum (1963) contains a great deal of significant new material.

———, 1973a: "The Ontology of the Curvature of Empty Space in the Geometrodynamics of Clifford and Wheeler," in Suppes (1973), pp. 268–295.

A critique of Wheeler's influential geometrodynamics.

Grünbaum, Adolf, Wesley C. Salmon, Bas C. van Fraassen, and Allen I. Janis, 1969: "A Panel Discussion of Simultaneity by Slow Clock Transport in the Special and General Theories of Relativity," *Philosophy of Science*, vol. 36, no. 1, pp. 1–81.

A detailed response to Ellis and Bowman (1967).

Hinckfuss, Ian, 1975: *The Existence of Space and Time*. Oxford: Clarendon Press.

An introductory survey of problems in the philosophy of space and time, focusing on the controversy between "absolutist" and "relationist" viewpoints.

Kamlah, Andreas, 1979: "Hans Reichenbach's Relativity of Geometry," in Wesley C. Salmon, ed., *Hans Reichenbach: Logical Empiricist* (Dordrecht: D. Reidel Publishing Co.), pp. 251–266.

A historical study of the origins of Reichenbach's views on space and geometry.

Machamer, Peter K., and Robert G. Turnbull, eds., 1976: *Motion and Time, Space and Matter*. Columbus: Ohio State University Press.

A collection of philosophical and historical essays; the subtitle of the volume is "Interrelations in the History of Philosophy and Science."

Malament, David, 1977: "Causal Theories of Time and the Conventionality of Simultaneity," *Nous*, vol. 11, no. 3, pp. 293–300.

A demonstration of the unique causal definability of the relation of simultaneity; a provocative result that *may* undermine the conventionality thesis of Reichenbach, Grünbaum, and myself.

Nagel, Ernest, 1961: *The Structure of Science*. New York: Harcourt, Brace, & World.

Chapters 8 and 9 challenge Reichenbach's views on equivalent descriptions and universal forces.

Nerlich, Graham, 1976: *The Shape of Space*. Cambridge: Cambridge University Press.

A lively defense of an anti-conventionalist theory of space and geometry.

———, 1979: "Is Curvature Intrinsic to Physical Space?" *Philosophy of Science*, vol. 46, no. 3, pp. 439–458.

A rebuttal to Salmon (1977a).

———, forthcoming: "Simultaneity and Convention in Special Relativity," in Robert McLaughlin, ed., *What? Where? When? Why?: Essays on Induction, Space and Time, and Explanation*.

Poincaré, Henri, 1952: *Science and Hypothesis*. New York: Dover Publications.

Written in French at the beginning of the 20th century, this classic statement of the conventionalist viewpoint was first published in English in 1905.

Prokhovnik, S. J., 1967: *The Logic of Special Relativity*. Cambridge: Cambridge University Press. Second edition, Kensington, N.S.W.: New South Wales University Press, 1978.

A nonstandard approach to special relativity on the basis of a "substratum theory." The second edition discusses the bearing of cosmic background radiation upon these issues.

Putnam, Hilary, 1963: "An Examination of Grünbaum's Philosophy of Geometry," in Bernard H. Baumrin, ed., *Philosophy of Science: The Delaware Seminar* (New York: John Wiley & Sons), pp. 205–215.

A spirited attack on Grünbaum's (1962); it is answered in Grünbaum (1968, chap. III).

Reichenbach, Hans, 1920: *Relativitätstheorie und Erkenntnis Apriori*. Berlin: Springer.

A classic attack on Kant's thesis of the synthetic a priori

character of geometry; English translation, Reichenbach (1965).

———, 1924: *Axiomatik der Relativistischen Raum-Zeit-Lehre*. Braunschweig: Vieweg.
A classic attempt to axiomatize relativity theory; English translation, Reichenbach (1969). Winnie (1977) contains some serious criticisms.

———, 1928: *Philosophie der Raum-Zeit-Lehre*. Berlin & Leipzig: Walter de Gruyter.
This classic treatise is the fountainhead from which most serious work on the philosophy of space and time in the 20th century has, directly or indirectly, sprung. English translation, Reichenbach (1958).

———, 1949: "The Philosophical Significance of the Theory of Relativity," in Paul A. Schilpp, ed., *Albert Einstein: Philosopher-Scientist* (Evanston, Ill.: The Library of Living Philosophers), pp. 287–312.
"I can hardly think of anything more stimulating as the basis for discussion in an epistemological seminar . . . ," said Einstein regarding this essay. See Einstein's discussion, *ibid*., pp. 676–679.

———, 1951: *The Rise of Scientific Philosophy*. Berkeley: University of California Press.
Chapters 8 and 9 contain very elementary discussions of time and space respectively.

———, 1956: *The Direction of Time*. Berkeley: University of California Press.
An insufficiently appreciated gold mine of discussions on the nature of time.

———, 1958: *The Philosophy of Space and Time*. New York: Dover Publications.
The outstanding 20th-century classic in the philosophy of space and time; a seminal and highly influential work. Translation of Reichenbach (1928).

———, 1965: *The Theory of Relativity and A Priori Knowledge*. Berkeley: University of California Press.
This book is Reichenbach's original attack upon Kant's theory of space, time, and causality. Translation of Reichenbach (1920).

———, 1969: *Axiomatization of the Theory of Relativity*. Berkeley: University of California Press.
Reichenbach's attempt to axiomatize the special and general theories of relativity; translation of Reichenbach (1924). Important criticisms of this axiomatization are given in Winnie (1977).

Robb, A. A., 1914: *A Theory of Time and Space*. Cambridge: Cambridge University Press.

———, 1921: *The Absolute Relations of Time and Space*. Cambridge: Cambridge University Press.

———, 1936: *The Geometry of Space and Time*. Cambridge: Cambridge University Press.

> These three books by Robb constitute an important, and generally neglected, axiomatic approach to relativity theory. See Winnie (1977) for a clear exposition of a spacetime theory based upon Robb's work.

Russell, Bertrand, 1956: *An Essay on the Foundations of Geometry*. New York: Dover Publications.

> Originally published in 1897, this work constitutes an attempt to defend Kantian a priorism in the face of non-Euclidean geometry.

Salmon, Wesley C., ed., 1970: *Zeno's Paradoxes*. Indianapolis: The Bobbs-Merrill Co.

> An anthology of important modern discussion of Zeno's paradoxes. Contains an extensive bibliography.

———, 1977: "The Philosophical Significance of the One-Way Speed of Light," *Nous*, vol. 11, no. 3, pp. 253–292.

> A general survey of issues relating to the conventionality of simultaneity, including detailed discussions of many putative methods for ascertaining empirically the one-way speed of light. Written at an introductory level for philosophical readers who have little or no background in philosophy of science.

———, 1977a: "The Curvature of Physical Space," in Earman, et al. (1977), pp. 281–302.

> A defense of the thesis that physical space has no intrinsic curvature, with special attention to affine curvature. An answer to Glymour (1972). Grünbaum offers a different answer to Glymour in Grünbaum(1973), pp. 773–788.

Saunders, John, and John Norton, forthcoming: "Einstein, Light Signals and the ϵ-decision," in Robert McLaughlin, ed., *What? Where? When? Why?: Essays on Induction, Space and Time, and Explanation*.

> A critical historical/philosophical discussion of the conventionality of simultaneity.

Schlesinger, George, 1964: "It Is False That Overnight Everything Has Doubled in Size," *Philosophical Studies*, vol. 15, pp. 65–71.

> An attack on the view that it is meaningless to assert that everything in the universe has doubled in size. Grünbaum (1968, chap. II) contains a response to Schlesinger's argument.

———, 1967: "What Does the Denial of Absolute Space Mean?" *Australasian Journal of Philosophy*, vol. 45, pp. 44–60.

> Further development of arguments in Schlesinger (1964); also answered by Grünbaum in (1968, chap. II).

Sklar, Lawrence, 1974: *Space, Time, and Spacetime*. Berkeley: University of California Press.
  A general survey of the philosophy of space and time, more detailed and more advanced than the present book.

——, 1979: "What Might Be Right about the Causal Theory of Time," in Wesley C. Salmon, ed., *Hans Reichenbach: Logical Empiricist* (Dordrecht: D. Reidel Publishing Co.), pp. 367–384.
  A critical discussion of Reichenbach's causal theory of time.

Smart, J. J. C., ed., 1964: *Problems of Space and Time*. New York: The Macmillan Co.
  An anthology of readings, contemporary and historical, on the philosophy of space and time.

Suppes, Patrick, ed., 1973: *Space, Time and Geometry*. Dordrecht: D. Reidel Publishing Co.
  A collection of contemporary essays, most of them relatively technical. The purpose of the volume, according to the editor, is to encourage greater formal rigor in the philosophy of space and time.

Swinburne, Richard, 1968: *Space and Time*. London: Macmillan.
  An approach to the philosophical problems of space and time by way of linguistic analysis.

Törnebohm, Håkan, 1963: *Concepts and Principles in the Space-Time Theory within Einstein's Special Theory of Relativity*. Gothenburg: Gothenburg Studies in Philosophy.
  A technical study in the foundations of special relativity.

Torretti, Roberto, 1978: *Philosophy of Geometry from Riemann to Poincaré*. Dordrecht: D. Reidel Publishing Co.
  Historical treatment of philosophical developments in the second half of the 19th century.

van Fraassen, Bas C., 1970: *An Introduction to the Philosophy of Time and Space*. New York: Random House.
  An introductory text, somewhat more technical than the present book, with a wealth of interesting historical material.

Wheeler, John Archibald, 1962: "Curved Empty Space-Time as the Building Material of the Physical World: An Assessment," in Ernest Nagel, Patrick Suppes, and Alfred Tarski, eds., *Logic, Methodology and Philosophy of Science* (Stanford: Stanford University Press), pp. 361–374.
  A brief nontechnical survey of the basic ideas of geometrodynamics. Grünbaum (1973a) and Salmon (1977a) are criticisms of the views expressed by Wheeler in this essay. Graves (1971) defends Wheeler.

Whiteman, Michael, 1967: *Philosophy of Space and Time*. London: George Allen & Unwin Ltd.
  A phenomenological approach to the philosophy of space and time, in sharp conflict with the approach of Reichenbach and Grunbaum.

Winnie, John, 1970: "Special Relativity without One-Way Velocity Assumptions," *Philosophy of Science*, vol. 37, nos. 1–2, pp. 81–99, 223–238.
   An ingenious defense of the conventionality of simultaneity.

———, 1977: "The Causal Theory of Space-Time," in Earman, et al. (1977), pp. 134–205.
   Development of the four-dimensional Minkowski spacetime on the basis of causal relations alone. This approach employs ideas originated by Robb in (1914, 1921, and 1936).

———, ed., 1977a: "Symposium on Space and Time," *Nous*, vol. 11, no. 3.
   A collection of articles by John Earman, Clark Glymour, Wesley Salmon, and David Malament on current issues in the philosophy of space and time. In his editorial introduction, Winnie comments on the relationships between the papers by Salmon and Malament, which appear, at least superficially, to be in sharp conflict with each other.

INDEX

Aberration of starlight, 112-13
d'Abro, A., 127
Absolute motion, 129
Absolute rest, 94
Absolute simultaneity, 81
Absolute space, 23, 59, 129
Absolute time, 129
Academy (Plato's), 2
Acceleration, 48, 63, 90, 95-97, 119, 122, 137
   infinite, 48, 51
Accelerator, particle, 117; *see also*, Stanford Linear Accelerator
Achilles and the Tortoise, paradox of, 32-38, 43-48, 51-53; *see also* Runner; legato, staccato
Addition, definition of, 54; *see also* Sum
Additivity, countable or denumerable, 54
Adler, Irving, 29, 131
Aging, human, 93
Airliner, jet, 93
Aleksandrov, A. D., 29, 131, 135
Aleph null, 53
American Association of Physics Teachers, 127
Amorphousness, metrical, 59, 135; *see also* Metric, intrinsic
Analyst, The, 39
Analytic statement, 17, 132
Angles, sum of, *see* Sum
Anomaly, causal, 123
A posteriori, vs. a priori, 21
Applied geometry, *see* Geometry
Approximation, 66
A priori arguments, 21, 28, 63-66; *see also* Synthetic a priori
Archimedes, 128
Aristotle, 3, 31-32, 36, 53
Arithmetization of calculus, 62
Arrow, 64-65
   paradox of, 33-35, 38-42
Asquith, Peter, 135
Astronauts, 93

Astronomy, 130
   history of, 69
At-at theory of motion, 41-42
Atom:
   material, 52
   of motion, 64
   spatial, 34-35, 64-66
   temporal, 34-35, 64-66
Atomic clock, *see* Clock
Austin, A.K., 49, 51-52
Axioms, 2, 131

Bell, 71
Benacerraf, Paul, 134
Benardete, José, 67
Bergson, Henri, 39, 53, 64, 133, 135
Berkeley, George, 39
Binary notation, 58
Bird, 49, 77-78, 136
Black holes, 130
Black, Max, 43-48, 51, 134
Bolyai, Johann, 6, 10, 14, 29, 131
Bondi, H., 138
Bonola, Roberto, 29
Born, Max, 90, 127, 135, 140
Bowman, Peter, 109-110, 115, 125-26, 139
Boy-girl-dog problem, 49-52
Boyer, Carl B., 67, 133
Bradley, F. H., 63
Bridgman, P.W., 63, 109-110, 127, 135, 139
Bright spot, Poisson, 69
Brush, Stephen G., 136
Burkill, J. C., 134

c (cardinal number), 55-58
   speed of light, *see* Light
Calculus, infinitesimal, 35, 39, 42, 62-63, 128-29, 131
Cantor, Georg, 24, 53, 55-56, 58
Cardinal number, *see* Aleph, c
Carnap, Rudolf, 132
Carroll, Lewis, 49
Cauchy, A., 35-36, 39-40

Causality, 105, 121-22, 126
Charge, quantization of, 63
Circularity (logical), 105
Circumnavigation, 93
Classical physics, 69, 71, 81, 90, 105, 110, 115; *see also* Newtonian physics
Clock:
  atomic, 93, 137
  grandfather, 140
  light, 82-87, 111-12
  macroscopic, 93-94
  microscopic, 94
  moving, 84-85
  transported, 96-100, 107-110
Clock paradox, 85, 93-100, 107, 117-18, 137
Common notions, 131
Composition, fallacy of, 134
Congruence, 23-28, 58-62, 133
  definition of, 23-24, 58, 62
Consistency, 14, 16
Continuity, 24, 35, 62-63, 66, 133, 135
Continuum, 42, 45, 52-53, 57, 59, 62
  cardinality of, 55-58
Contraction, length, 85, 87, 100, 114-19, 136-37
  symmetry of, 87
Contradiction, *see* Inconsistency
Convention, 59-62, 101-104, 110, 113-27 passim
  non-triviality of, 110
  see also Definition
Convergence, 35-38, 47, 53-55, 135
Cooper, Leon, 140
Coordinates, 7, 54-55, 58-62, 135
Coordinating (coordinative) definition, *see* Definition
Copernicus, 69
Corpuscular theory (light), 70
Correspondence, one-to-one, 53, 56, 62
Countable infinity, 53
Courant, Richard, 67, 135
Cramér, H., 134
Curvature, 11-21, passim
  constant, 131
  external, 132
  Gaussian, 132
  internal, 132, 135
  mean, 132
  negative, 11
  positive, 11
  variable, 131
Curvilinear figure, 16
Cylinder, 132

Deceleration (tachyons), 122
Dedekind, R., 35
Definition, 45, 101, 125
  of congruence, 23-24, 58, 62
  coordinating (coordinative), 27-28, 61, 105
  geometry chosen by, 26
  of simultaneity, 73, 104-105, 109-110, 113, 115, 136
Dense order, 42
Denumerable infinity, 53
Derivative, 35, 38-39, 41, 48, 63
Descartes, R., 2-3, 128
Descriptions:
  equivalent, *see* Equivalent Descriptions
  simple, 106
Diagram, space-time, 76-80, 124
Dichotomy paradox, 32-53 passim
Differentiability, 63
Differential effect, *see* Effect
Differential force, *see* Force
Dilation, time, 81-87, 93-97, 100, 110-11, 115, 117, 136
  symmetry of, 85, 87
Dilemma, 52-54
Discontinuity, 51, 66
Discontinuum, Cantor's, 56-58
Discreteness, 42, 64, 66, 133
Distance, 55, 59-60; *see also* Length, Measure theory
Divisibility, infinite, 52
Doppelganger, 47

Earth, orbital motion of, 71, 109
Edwards, Paul, 133
Effects:
  differential, 22-23, 132
  universal, 23, 132
Egyptian geometry, 1-2, 4
Einstein, Albert, 25, 45, 69-91, 93-119 passim, 125-129, 135-36, 138
  Autobiography, 72, 100
Electromagnetic theory, 63, 69-73, 100
Electron, 90
Elements (Euclid), 2, 5, 29
Ellis, Brian, 109-110, 115, 125-26, 139
Embedding, 17-18, 132

INDEX

Epsilon (ϵ):
  choice of, 106
  ϵ-Lorentz transformations, *see*
    Lorentz transformations;
  non-standard value, 115
Equal passage time, principle of, 116
Equivalent descriptions, 25, 61, 106, 125, 133
Ether, luminiferous, 71-72, 84
Euclid, 2, 4-6, 14, 29, 128, 131
Euclidean geometry, *see* Geometry
Extension, 52
External curvature, *see* Curvature
External visualization, *see* Visualization

Fact, 101-106, 113; *see also* Convention
Factual content, 115, 127
Factual core, 116-17
Feinberg, Gerald, 140
Fine structure of space-time, 64
Finite point set, 54
First signal, 105, 111, 124
Fizeau, Armand, 101-103, 106, 113
Force:
  differential, 22, 25, 58, 132;
  universal, 16, 22, 25, 58, 61-62, 132
Four-dimensional space, 17-18
Frame of reference, 72-90, 96-100, 125
  privileged, 72, 84
Friedberg, Richard, 48, 51
Function, mathematical, 35, 39-43, 46, 63, 135
  limit of, 42

Galileo, 69
Gardner, Martin, 48-50, 52
Gauss, Carl Friedrich, 6, 15, 22
Geocentric system, 69
Geodesic, 7
Geometrical paradox, 52
Geometry: analytic, 3, 62, 128
  applied, 28
  differential, 131
  Euclidean, 2-29 passim, 129, 131-32
  non-Euclidean, *see* Non-Euclidean geometry
  pure, 27-28
Gravity, 94, 105, 130
Great circle, 7, 10

Grünbaum, Adolf, 29, 47, 51, 54, 59, 67, 96, 110, 125, 132, 133-35, 138

Hafele, J. C., 93-94, 137
Hal, 43-48 passim
Halsbury, Lord, 96, 98, 115, 117
Hartshorne, Charles, 133
Heath, Sir Thomas L., 29, 131
Heliocentric system, 69
Helmholtz, Hermann von, 18-20
Hertz, H., 69
Hilbert, David, 29, 131
Holton, Gerald, 136
Homogeneity, 59
l'Hospital's rule, 109
Hump, world with, 21-25, 131

Idealization, 45, 63
Illusion, 3, 87
Inconsistency, 94-96, 132
Indirect proof, 5
Inertial frame, 72, 78, 82, 87, 95-96, 114-117, 125, 136; *see also* Frame of reference
Infinite sequence, *see* Sequence
Infinite series, *see* Series
Infinitesimal calculus, *see* Calculus
Infinitesimal distances and times, 39
Infinity machine, 43-52, 134
Information, transmission of, 139
Initial path, 50
Instant, 34-35
Integral, 35
Internal visualization, *see* Visualization
Interval:
  on line, 54
  space-time, 122-24
Intrinsic metric, *see* Metric, intrinsic
Intuition, spatial, 17, 20, 43; *see also* Visualization
Invariant, 121-22, 126, 140
Irrational points, measure of, 55
Isomorphism, 24, 56

Janis, Allen I., 139
Jupiter, 73, 106-108

Kant, Immanuel, 4, 6, 16-17, 20, 27, 129, 132
Keating, Richard E., 93, 137
Kepler, J., 69

Kolmogorov, A. N., 29, 131, 135
Kreisler, Michael N., 139

Lamp, *see* Thomson lamp
Latitude, 7-8
Lavrent'ev, M. A., 29, 131, 135
Lee, H. D. P., 134
Legato runner, 47-48
Leibniz, G. W., 2, 128-29
Length:
   of interval, 54-55, 61-62
   of moving object, 85, 88, 115-16
Light, 69, 71, 82
   clock, *see* Clock;
   constancy of speed, 70, 82, 101, 103, 106, 111, 113
   faster than, *see* Pseudo-process, Super-light, Tachyon
   finitude of speed, 106
   first signal, 110-111, 124
   measurement of one-way speed, 104-107, 110-113, 126
   measurement of speed, 101-103, 106, 108
   one-way principle, 101, 103, 106, 113-14
   one-way speed, 105, 107, 110-13, 117-19, 126
   path of ray, 15-16, 21, 28, 73, 78
   round-trip speed, 103-104
   signal, 100, 121
   speed, 71-75, 83, 90, 95, 100-101, 105, 110, 136
   two-way principle, 101, 103, 113, 116
   wavelength, 133
Light-meter, 78, 82, 85, 136-37
Light year, 136-37
Lightning, 73-80, 104
Limit, 35-37, 42-43, 46, 134
Line, 52-53; *see also* Straight line
Linearity principle, 116-17
Lobachevski, Nikolai Ivanovich, 6, 10, 14, 19, 29, 131-32
Locomotive, 49
Logic, 31
Longitude, 7-8, 10
Lorentz transformations, 85, 87, 107, 114-19, 136
   ϵ-Lorentz, 116-19
Lovejoy, A. O., 137

Machine, *see* Infinity machine
Magnitude, zero, 52

Mathematics:
   abstract system, 27
   pure vs. applied, 45
Maxwell, J. C., 69-70, 72-73, 90, 100
Measure theory, 54-59, 128, 134
Measuring rod, 16, 20-24, 28, 59, 61-62, 133
Meditations (Descartes), 3
Medium (for waves), 70-71
Mercury, precession of perihelion, 129
Meter (standard), 133
Metric, intrinsic, 24, 59, 132-33, 135
Metric rule, 60-62, 135
Michelson, A. A., 71, 103
Michelson-Morley experiment, 71, 103, 113, 136
Mill, John Stuart, 4
Minkowski diagram, *see* Diagram, space-time
Montague, W. P., 137
Morley, E. W., 71
Mossel, M., 139
Motion:
   functional relation, 40
   instantaneous, 38-39
   Newton's first law, 117, 136
   paradoxes of, 31-52
   uniform, 116-17
Mountaintop, 15-16
Muon (mu meson), 94, 115, 117

Nanosecond, 93
Newman, James R., 133
Newton, Isaac, 2, 23, 59, 69, 90, 104-105, 117, 119, 122-29 passim
Newton, Roger G., 140
Newtonian physics, 69, 104-105, 119-26, passim
Nextness, 42
Non-denumerable sets, 53-55
Non-Euclidean geometry, 4-28 passim, 129
Non-Euclidean space, 17-20, 132
Non-Euclidean surfaces, 17-18
Non-measurable sets, 55
Non-triviality (of conventions), 127, 140
Numbers:
   infinite, 53
   rational, 55
   real, 35, 62

Observable consequence, 116-17, 119
Observation, inanimate instruments

INDEX

of, 76
Observatory, U.S. Naval, 93
Observer, 72-90 passim
One-way speed, 103-104; *see also* Light, one-way speed
Operation, physical vs. mathematical, 45
Ordering, internal, 55
Owen, G. E. L., 133

Pal, 43-48 passim
Parallax, stellar, 112, 132
Parallels, 5-10
Parity, logical vs. epistemological, 17, 20
Parmenides, 31
Partial sums, sequence of, 37-38, 53
Parts:
   non-existence of, 31
   of instant, 34
   ultimate, 53
Passage times, 116-17
Peirce, Charles Sanders, 35, 51, 133
Perturbation, *see* Force, universal; Effect, universal
Photon, 137
Physical reality, 46, 62
Physical space, 15-28 passim, 58, 62, 66
Pittsburgh Panel, 110
Plane, Euclidean, 7-8, 132
Plato, 2-3, 27
Playfair's Axiom, 131
Plurality, paradox of, 24, 31-32, 35, 52-58, 63
Poincaré, Henri, 16, 21-23, 25, 29, 131
   two-dimensional, 22-25
Poisson, S. D., 69
Poisson bright spot, 70
Pollard, Art, 137
Postulates:
   of geometry, 2, 4-6, 27, 131
   of special relativity, 72, 82, 116
Potato chip, 13
Progressive dichotomy paradox, *see* Dichotomy
Projection, simultaneity, 90, 115, 118
Proof, 2, 27, 128
Proportional passage times, principle of, 117
Pseudo-processes, 112, 139
Pseudo-sphere, 11, 14, 131-32
Ptolemy, 69

Pure geometry, *see* Geometry
Pythagoras, 2, 65-66, 82, 128
Pythagorean theorem, 65-66, 82

Quantization, 63-66, 135

Ratio, circumference to diameter, 10-14, 18-21
Real numbers, *see* Numbers
Reality:
   physical, *see* Physical reality
   ultimate, 3, 31
Reciprocity condition, 86-87, 114, 116, 140
Red-shift, gravitational, 129
Reductio ad absurdum, 5, 69
Regress, infinite, 42
Regressive Dichotomy paradox, *see* Dichotomy
Reichenbach, Hans, 21, 23-24, 29, 105-106, 111-12, 114-15, 124-25, 127, 132-33, 139
Relativity:
   factual content, 119
   general, 72, 94, 96-97, 129-30
   principle of, 84
   special, 45, 69-127, 129-30, 135-36
   testable consequences, 90, 115
Rest, instantaneous, 38
Retardation of clock, 84, 94, 98, 100, 107, 109-110, 115, 117
Riemann, Georg Friedrich Bernhard, 6-7, 14, 131
Riemannian geometry, 7
Robbins, Herbert, 67, 135
Rocket ship, 90, 93-94, 122
Rod, *see* Measuring rod
Römer, Olaf, 106-110, 119, 126, 139
Rotation, 129
Round-trip speed, *see* Light: round-trip speed
Runner, legato vs. staccato, 47-48
Runs, infinite sequence, 43, 46
Russell, Bertrand, 2, 31-32, 34, 41, 133-34

Saccheri, Girolamo, 5-6
Saddle surface, 11-14, 21, 131
Salmon, Wesley C., 67, 110, 133-35, 139
Schilpp, P. A., 136, 138
Self-evident propositions, 4-5
Separation, *see* Interval
Sequence, infinite, 35-36, 53

Series:
   convergent, 36-38, 47
   infinite, 35, 53
   sum, 36, 43, 45
Set theory, 24, 53
Shankland, R. S., 112, 139
Signals:
   arbitrarily fast, 122;
   first, 110-111, 124;
   super-light, 104, 119-22;
   see also Light
Similar triangles, 5
Simultaneity, 73-81, 88-90, 97-99, 112, 119-27, 129, 139-40
   absolute, 119, 122
   conventionality of, 114, 125, 127
   definition of, 73, 104-105, 109-110, 113, 115, 136
   distant, 73, 107, 109, 136
   local, 73
   projection, 90, 115, 118
   relativity of, 76, 85, 87, 100, 114-15, 122, 124-25, 136
   see also Synchrony
Smart, J. J. C., 29
Sound, 70-71
   speed, of, 104
Space-atom, *see* Atom, spatial
Specious present, 42
Sphere, 7-12, 18-19, 21, 132
   in three-dimensional space, 19
Staccato runner, 47-48
Stadium paradox, 34-35, 64
Stanford Linear Accelerator, 90, 115
Starlight:
   aberration, 112-13
   bending, 129
Stedman, G. E., 139
Stipulation, *see* Convention
Straight line, 4-12, 16, 26-28
Subjectivity, 76
Sum:
   angles of triangle, 5-21 passim
   of infinite series, 37, 53-55
   of zeroes, 54-56
Super-light signal, *see* Signal
Supertask, 44-45, 50, 134
Swayze, John Cameron, 95, 137-38
Switch, *see* Thomson lamp
Synchrony, 96-100, 105-110, 115-21, 139
   absolute, 122
   clock transport, 126
   local, 96, 107, 115, 117
   non-standard, 114
   slow clock transport, 109-110, 119, 125-26
   standard signal, 106, 109-110, 114, 119
   see also Simultaneity
Synthetic:
   a priori, 4, 16, 20, 27, 129
   vs. analytic 17
System, *see* Frame of reference

Tachyon, 105, 122-25, 139
Tasks, infinite sequence, *see* Supertask
Taylor, Edwin F., 91, 127, 136-37, 140
Telescope, moving, 112-13
Temperature, changes, 22
Terms, primitive, 27
Ternary notation, 57
Thales of Miletus, 1-3, 31, 128
Theorem, 2
Thomson, James, 45, 134
Thomson lamp, 44-48, 51, 134
Three-dimensional space, 14-20 passim
Tiles, space composed of, 65-66
Time-atom, *see* Atom, temporal
Time:
   proper, 97
   psychological, 42
   see also Simultaneity; Synchrony
Timex watches, 93, 137-38
Toni home permanent, 138
Train, Einstein's, 69-90, 136
Transferring machine, *see* Hal; Pal
Transmission of information, 105, 120; *see also* Signal
Triangle:
   Gauss's, 15
   sum of angles, *see* Sum
Trojan fly, 49-51
Tunnel, 87-88
Twin paradox, 81, 84, 93-96, 118, 137-38
Twin-rod experiment, 118
Two-dimensional surface, 6-7, 17-21

Uncountable infinity, 53
Undetectability, empirical, 23
Uniform motion, *see* Motion
Units of measure, 26, 78, 82, 84, 136
Universal effect, *see* Effect

INDEX

Universal force, *see* Force

Vacuum, speed of light in, 71
van Fraassen, Bas C., 29, 110, 127, 139
Variable, epsilon as, 116
Veblen, Oswald, 29, 131
Velocity:
   arbitrarily high, 47, 105, 119
   average, limit of, 39
   bastard, 109
   composition of, 45, 90, 99, 122
   discontinuity, 48, 51
   function continuous, 48, 63
   instantaneous, 38-41
   intervening, 110
   limiting, 105, 109
   one-way, 109, 116-17, 120
   self-measured, 109-110
   standard (discrete space and time), 64
   see also Light

Visualization, 17-20, 132
Vlastos, Gregory, 133

Wave, 69-72
Weierstrass, K. T., 35, 41
Weinberger, H., 139
Weiss, Paul, 133
Weyl, Hermann, 65-66, 134-35
Wheeler, John Archibald, 91, 127, 136-37, 140
Whitehead, Alfred North, 64
Winnie, John, 115-18, 125, 139-40
Wisdom, J. O. 133
Wolfe, Harold E., 29
Woods, Frederick S., 29
World line, 77-80

Young, J.W.A., 29, 131

Zeno of Elea, 31, 128, 133
Zeno's paradoxes, 24, 31-67, 133
Zeroes, addition of, 58